The Practical Guide to Wwoofing 2012

by A. Greenman

A delightful and informative read, very accessibly and humanly written- it almost reads itself to you!

Sue Coppard - Founder of Wwoof

"A fantastic guide book"
Wwoof Australia

**The Practical Guide
to Wwoofing 2012**
By A. Greenman

Copyright ©2012 A. Greenman

6th Edition
Published by Greenmans Books
in association with Lulu Press

www.greenmansbooks.com

First Published in 2009
by Lulu Press

ISBN - 978-1-4709-3574-0

Please note that this book is not written by W.w.o.o.f and that the opinions expressed here do not necessarily represent those of any Wwoof organization.

W.W.O.O.F

WORLD WIDE OPPORTUNITIES/WILLING WORKERS ON ORGANIC FARMS

"A wonderful book"
Wwoof Canada

Other titles by A. Greenman:

How to Take a Gap Life - *Sometimes a Gap Year is Just Not Enough!*
The Wisdom of Travel - *Words to Inspire*
A. Greenman's Short Stories

The Adventures of a Greenman Series:

Part 1 - *The First 30 Years of a Greenman*
Part 2 - Raw Travel Brazil
Part 3 - Raw Travel France
Part 4 - Raw Travel Spain
Part 5 - Raw Travel India
Part 6 - Raw Travel New Zealand
Part 7 - Raw Travel Wales
Part 8 - Raw Travel Eastern Europe
Part 9 - Raw Travel Hungary
Part 10- Raw Travel England
Part 11 -Raw Travel Crete
Part 12 -Raw Travel Italy
Part 13 -Raw Travel Europe
Part 14 -Around the World in a Decade
Part 15 -I Travel Light:
The Man Who Walked Out of the World.

www.greenmansbooks.com

CHAPTERS - PAGE NUMBERS

1 Questions and Answers - *9*

'Wwoof', 'Wwoofers' and 'Wwoofing'

2 Countries you can Wwoof in - *19*

3 About the Host List - *23*

4 Duration of Stay - *25*

5 The Work Itself - *27*

6 Safety - *33*

7 Livestock - *37*

8 Accommodation - *41*

9 Meals, Food and Diet - *43*

10 The Setting - *47*

11 Money - *51*

12 Language - *55*

13 Sex'n'Drugs'n' Rock 'n' Roll - *59*

14 Arranging Your Stay - *63*

15 Travelling to Hosts - *67*

16 Cycling - *71*

17 On foot - *73*

18 Hitchhiking - *77*

19 Baggage - *81*

20 Equipment Needed - *85*

21 In between wwoofing - *91*

22 Joining Wwoof - *97*

23 Visas - *115*

24 Returning Home - *119*

25 More Questions and Answers - *121*

26 A Wwoof Journey in England - *127*

*

WWOOFS MAIN AIMS ARE FOR PEOPLE TO:
GET INTO THE COUNTRYSIDE
HELP THE ORGANIC MOVEMENT

GET FIRST HAND EXPERIENCE OF ORGANIC
FARMING AND GROWING

MAKE CONTACT WITH OTHER PEOPLE IN THE
ORGANIC MOVEMENT

Take my hand and rest a while.
Hold on tight and I will take you on a journey,
Into the world of Wwoofing.

There is space between the pages
To record your thoughts for ages.

CHAPTER 1

QUESTIONS AND ANSWERS

Where, when and why did 'wwoof' start?

Wwoof started in London, England in 1971. To put people in contact with places. People who wished to help on organic farms. Farms that would like volunteers.

Who started it?

'Sue Coppard' founded wwoof, originally known as 'weekend workers on organic farms'. As the movement became more popular and people began to volunteer during weekdays, the name changed to 'Willing Workers on Organic Farms'. This name remained for many years. Other autonomous organizations arose in other countries and Wwoof became 'World Wide Opportunities on Organic Farms' (though some still keep the 'Willing Workers' title.)

Is it only for people who grow carrots and talk about green things?

No. Many other consumables may begin life on organic farms. Dairy products, meat and alcohol for example. Hand crafted items such as soaps, essential oils and medicines also often have their origins on organic farms.

But why would someone not get paid anything work for nothing? I heard that the workers do.

Neither are they often required to be skilled, so the potential for learning is enormous. Or pay for food, accommodation or for the volunteering post itself. Many other schemes charge substantial amounts for this.

What is a 'wwoofer'?

Someone who has joined wwoof and is volunteering (wwoofing).

Does wwoof arrange work for you?

No. You arrange your stay, by contacting the farm. More about that later.

Then what do wwoof actually do?

Act as an intermediary. People who would like to have volunteers join Wwoof. Potential volunteers also join wwoof , for a small fee. Wwoof then issue a list of the people who are seeking help to the people who would like to help. Great idea eh!

Does one membership cover all the countries that belong?

No. The membership covers that country, for example Italy. Or a set of in the case of the United Kingdom, e.g. England, Scotland, Wales. Some countries club together and you get details of all opportunities there. More about that later in chapter 22 'Joining'.

How many countries can I wwoof in at the moment

97. The number is growing.

Will the points raised in 'The Practical Guide to Wwoofing' apply to many countries?

Yes, almost certainly. This book is intended to cover many of the points that will crop up time and time again in countries throughout the world that you may wwoof in. Naturally, not everything will apply in all countries, however the chapters have been designed to help wwoofers of any nationality – regardless of which country they choose to wwoof in.

Do I need to send a CV to get a wwoofing position?

Usually not, a brief outline about yourself can be enough. If the member thinks you will fit in and they have space at your requested time, you just agree to go then. It you cannot attend – tell them beforehand.

Is it just farms that take people?

No. It is perhaps one of the most important points to recognise about members. There are also schools, educational centres, smallholdings, people with large gardens, country estates, communities and other places that receive volunteers. They are known as 'hosts'. It is a term used often in this movement. You will get a list of 'hosts' when you join wwoof.

If we are a group, can we just get one membership and pass around the details for all?

'The Practical Guide to Wwoofing' is not written by wwoof. However, this type of sharing is not appreciated. Wwoof is a charity, at least Wwoof UK. Other countries statuses vary, but they are generally non-profit making. It does not help them if the list that they work hard to collate, regulate and update is given away for free. Where possible, join wwoof and encourage other suitable candidates to do so.

Does 'The Practical Guide to Wwoofing' provide a list of hosts or at least recommendations of places to go?

No. We do not include the host list in this book. Nor do we recommend places to go. We are all different. Wwoofers and hosts. We all have a unique experience in our meeting. What one likes, another may not. Furthermore, some places are exceptionally beautiful and blessed with wonderful people. They will not get much land work done if they receive an abundance of communication from wwoofers! Finding the ones that you really click with is part of the excitement of wwoofing.

Do I have to raise any sponsorship money before I go?

No, unlike many other volunteering projects you do not.

Will wwoofing cost me anything else?
Are there any hidden charges?

There are no hidden charges. You will need to cover the cost of your transport, more about that later. Its also worth having some basic equipment. Many of the items you may already have. There is a section on this too. And of course, the membership fee (The equivalent of £20 or less for a year, depending on the country).

Is there any way of progressing in the wwoof movement? Gain qualifications?

There is no specific grading system. Sometimes a host will advertise for 'experienced wwoofers'. That is your prize if you become one.

What sort of person wwoofs?

There is no 'absolute' on this. If you are fit and able, it is certainly a good start. An interest in organic food, gardening or farming is important. A desire to learn is valuable. An age at what to do this...? Somewhere between 18 and your last breathing day is good.

Are you saying a pensioner could wwoof?

I have seen it happen. Eventually they married the host!

But is there an average age?

I have not seen a survey on this, though I notice that people in their twenties are quite common. I have met plenty of 18 and 19 year olds, men and women in their 30's and a wide range of people who are 40 +. Average age – perhaps someone will discover this if they have to.

Can I wwoof even if I am not entirely fit, perhaps with a disability or impairment of some kind?

Some hosts have work available for you even if you are not entirely fit. Do not discount the possibility of wwoofing just because you are not in top shape or perhaps use a wheelchair. For example the production of organic jams may include a task that you can do – without needing to work out in the field doing strenuous tasks. Or perhaps you can work on a farm for a shorter period of time and come to some arrangement with a host. Contact one and talk to them. Your help may be greatly appreciated. Be honest with your host when you contact them, let them know what your level of fitness is.

Will all of the wwoof hosts practices always be organic?

No, but mostly. Hosts have to meet fairly tight criteria. It is likely that at least a major aspect of their lives or sometimes business reflect a truly organic path. As you will discover, they vary tremendously. Some will not be involved in any work that involves chemicals. For example, non-toxic paint will be used on buildings. Others will use materials that contain chemicals or whose production has had a significantly negative effect on the environment – plastics for example. If you can relax a little around these issues, it may be helpful. Consider that a polythene tunnel may produce an abundance of fresh organic salads, sold locally. Perhaps this offsets the original negative impact in some way, perhaps not. You have to find your own way and make your own decisions about what is right for you.

Is wwoofing for me?

You may only really know this once you have been to some places and volunteered, in your home country, overseas or both. Occasionally one meets wwoofers who are really just looking for a free place to stay. Are you interested in organic farming or even vegetable gardens? Or does woodland management catch your eye? Perhaps working with horses, pigs, chickens? There are many other aspects of wwoofing, though should you not find at least one that attracts you, then wwoofing may not be for you. Likewise, very occasionally there maybe a host somewhere in the world who exploits the good nature and willingness of their wwoofers by just seeing them as free labour. Fortunately this is rare, but with nearly 100 countries in the world offering opportunities, it is certainly a possibility. For example, decorating the inside of a house for 2 weeks, one may have very little opportunity to learn about organics or even the land. On the other hand, some wwoofers may be happy to do this job, if only to be in a stunning countryside setting. More about the work later. Please consider though, that many tasks that seem dull or unfair, are actually essential parts of leading a natural lifestyle. Collecting manure for rich compost or endless hours of weeding are fundamental for the cycles of the land to continue effectively, in this manner. Popular hosts vary tasks and spend time sharing some of their knowledge with wwoofers. A good example:

The Wood Pile

On a farm I was volunteering at, I noticed an extraordinarily beautiful firewood pile. Very well stacked and fine in its presentation. I commented upon it to the hosts and he told me a little about his last wwoofer.
"He wanted something just to get his teeth into and spent the week splitting up logs and stacking them like that. He enjoyed it immensely"

So here is an example of a wwoofer who preferred to do the same job, perhaps he knew where he stood with his work then. It suited him to just get on with it. Although there was little variation in the tasks. A host can interact during breaks, mealtimes and evenings. Just being around a host who is mindful of the land can be an education in itself.

Is there anything else I should know about wwoofing before I start?

Yes, probably a 1000 things, but most you will learn as you wwoof. Or some may become apparent to you as you read this book.

Great, I want to read the book, but first I want to join Wwoof. How do I do that?

Go to Chapter 22!

CHAPTER 2

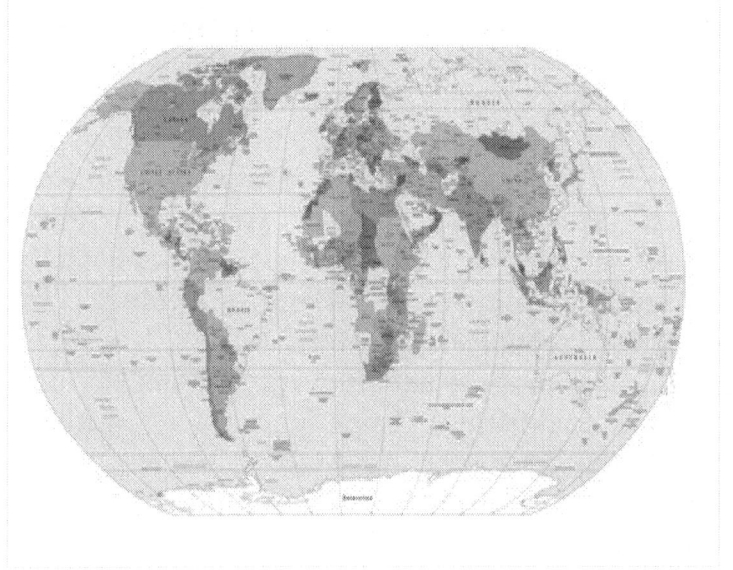

COUNTRIES YOU CAN WWOOF IN

Yes it is possible to wwoof yourself around the world!
All of the following countries have wwoofing opportunities:

Argentina, Australia, Austria, Bahamas, Bangladesh, Barbuda, Belgium, Belize, Benin, Bolivia, Brazil, British Virgin Islands, Bulgaria, Cameroon, Cambodia, Canada, Chile, Columbia, Cook Islands, Costa Rica, Croatia, Czech Republic, Denmark, Dominica (Commonwealth), Ecuador, Egypt, El Salvador, Estonia, Ethiopia, Finland, Fiji, France, French Polynesia, Gambia, Georgia, Germany, Ghana, Greece, Guatemala, Haiti, Hawaii, Holland, Honduras, Hungary, Iceland, Ireland, India, Indonesia, Ireland, Israel, Italy, Jordan, Kenya, Laos, Latvia, Lebanon, Luxembourg, Jamaica, Japan, Jordan, Kazakhstan, Kenya, Korea, Laos, Lithuania, Macedonia, Madagascar, Mali, Mongolia, Mexico, Moldova, Morocco, Mozambique, Namibia, Netherlands, Nepal, Nicaragua, Nigeria, Norway, New Zealand, Pakistan, Palestine, Panama, Peru, Philippines, Poland, Portugal, Senegal, Serbia, Sierra Leone, Slovenia, South Africa, Spain, Sri Lanka, Sweden, Switzerland, Taiwan, Tanzania, Thailand, Togo, Tonga, Trinidad and Tobago, Turkey, Uganda, Ukraine, United Kingdom, United States of America, Uruguay, Romania, Russia, Venezuela, Zambia, Zimbabwe.

Some countries have just a few farms where you can volunteer. Others have hundreds, even a thousand or more organic places where you can help and learn in exchange for your food and accommodation. What are you waiting for?

Gap Years

There are so many hosts around the world, it is possible to work your way around at least part of it. This may be an extremely worthwhile way of spending a 'gap year', or indeed a gap life as 'sometimes a gap year is just not enough!' You will be learning skills as you go that may serve you for life, going from one neighbouring country to the next. Of course, you can take time out from your gap life, earn a few shekels with work that interests you, in between wwoofing.

How useful to know about growing your own food, raising livestock and building your own shelter. In some cases, you may even learn how to make your own clothes, shoes or shampoo! Perhaps you have a useful skill yourself, that you can share with hosts. The possibilities with wwoofing are infinite. There are worse things to do with a year or two, or ten.

Later we will speak in much more detail about ways of funding your travels that need not cost the earth. One important question is how you will get from place to place, country to country. Are you happy with flying, taking cheap flights? Or using other forms of public transport? Or perhaps going under your own steam? You can spend months or even years planning such a trip, you can also just go!

*There is space between the pages
To record your thoughts for ages.*

CHAPTER 3

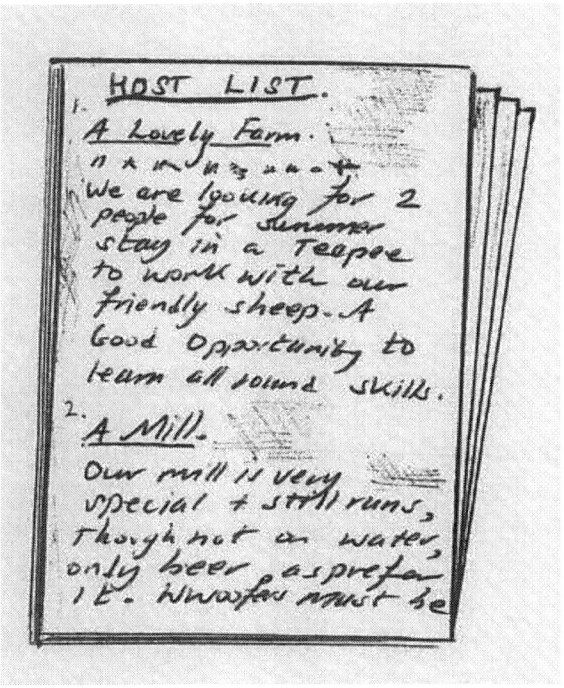

ABOUT THE HOST LIST

Once you have joined wwoof, the host list is sent along with your membership number. By post or by email. As well as a list of gardens, farms, smallholdings (smaller than a farm) and communities in your chosen country/or countries, it may include (amongst other things) the following information about each host:

Type of help needed; hours of help required; food/diet offered; accommodation provided; languages spoken, number of adults/children present; alcohol/smoking exclusions, some information on how to get there.

The information provided will vary from host to host, not only in length but also in the depth of its content. We will talk later a little more about contacting your hosts, also in much greater detail about the other points above. One may read the host list extensively, carry out much research in order to make your choice and still find the whole experience of arriving and spending the first few days at a place a little daunting, perhaps even questioning whether or not you have made the 'right' choice. One wwoofer I met just opened his host list at random and chose the host where his finger landed. He was quite happy with that and it seemed to work for him.

The host list may become your closest travelling companion, your 'wwoofing bible'. Hold onto it. It is the key to your journey and the source of much learning.

CHAPTER 4

DURATION OF STAY

The duration of your stay could be as little as 2 or 3 days. However, if this is possible it will be indicated in the host list. Commonly stays are 1-2 weeks. This gives the host and wwoofer a chance to see if they 'fit' together. If you get on well, you may be able to extend your stay or return another time. Though some hosts like wwoofers to stay a minimum of 2 months straight off. So again we see the variations that occur in the uniqueness of hosts criteria. Check this out with them before you go. Long term wwoofing in one place gives a real opportunity to integrate into a community. Farms may take you on for 6 months or a year or more once you have worked for an initial period. Perhaps we all find a place somewhere that feels like home.

Some long term wwoofers take part time work in a nearby town or village, others make things to sell. There are many ways to fund a journey that may effect the duration of your stay but it is more likely that this could happen if you are staying much longer with a host.
Of course it also needs to be okay with them that you undertake such part time work. Remember though, you should not get paid for wwoofing work or ask for money. It is an exchange and removing money from the equation adds much of the charm to the experience and can be one of the driving forces of the wwoof movement.

CHAPTER 5

THE WORK ITSELF

In Southern France.

We set out early from the farmhouse, accompanied by two donkeys. The farmer's wife, my self and another wwoofer head for a nearby forest. Our task was to collect twigs and fallen branches, make them into faggots (bundles) and ask the donkeys to bring them back down the mountain. I wondered if these lovely creatures would understand English. My host suggested that I did not under estimate the intelligence of our companions. Later they would be used to make a big fire in the bread oven (the branches). Once it was to an optimum temperature, the dough the farmer had prepared would go in. We would make 100 loaves and sell them in the village. All the ingredients were organic. Our delivery truck ran on a 50/50 mix of waste vegetable oil and diesel too.

Bread making would become a major part of my journeys, the style and method varying from country to country. Later, living in a sleepy village in the South of England, I found that the variations in bread making methods that I have observed whilst travelling came in very handy. Each morning I would make a batch of bread by hand and put them outside my front door for sale in a basket. In the evening I would bring in the basket (usually empty), then count the coins that have been placed in my honesty box. So wwoofing really can teach us new valuable life skills. What could be more so than providing your own bread and butter money?

Jobs you may get involved in could include:

Apple picking, Almond shelling
Bee keeping, Beetroot growing
Cider making, Chicken chasing
Dog walking, Dry stone walling
Elderberry wine making, Egg collecting
Felt making, Feeding cows
Grape squashing, Goat keeping
Horsemanship, Hay gathering
Igloo building, Ideas sharing
Jam making, Just about anything
Kangaroo spotting, Kicking footballs for host's kids,
Ladder climbing, Lambing
Market stall holding, Making friends
Navigating a Land Rover, Nettle collecting,
Orange squashing, Olive picking
Piglet itching, Planting wheat
Quilt making, Quinoa cooking,
Rat catching, Raspberry picking
Sowing seeds, Sawing wood
Tree house making, Thistle pulling
Under floor heating construction, Unbelievable jobs,
Venomous snake look out, Veranda sitting,
Weeding, Woodworking
Xactly everything you have ever dreamed of,
Yellow fin Tuna preservation, Yurt construction
Zoology assistant, 'Zzzzzzzzzz' a lot.

A Days Wwoofing in Scotland

7.45am Breakfast
8.15am Bottle feed orphaned goat kids and give hay to bullocks
9.00am Weeding in the poly tunnel
10.00am Coffee and cake
10.15am Dig a trench for potatoes
12.00pm Lesson on effective potato planting method
12.30pm Feed and play with the dog
1.00pm Lunch/ Wash up
2.00pm Go out with farmer to collect bales of hay
3.30pm Tea and Biscuits!
3.45pm Chop some firewood/light fire
4.30pm Hot Bath
5.30pm Watch television
6.30pm Help prepare dinner and eat
7.30pm Sew up my trousers
9.00pm Bed
10.00pm Dream of friendly bull

Timing

The wwoof world does not always start at 9.00am and finish at 5.00pm, so be prepared for some unusual hours. In Spain, it was hot so we worked from 9.00am to 2.00pm and then spent the afternoon on the beach. In very hot conditions I would sometimes start volunteering at 5.00am and finish at 11.00 am. Sometimes we would collect avocadoes in the evening or pick vegetables for the next day's market. Weekends were free.

The average number of hours I have worked per week is 30, however this can vary tremendously. One host asked for 6 hours a day, 3 days a week. Commonly you will work 5-6 days a week. Hosts are often very flexible and will help you to juggle hours around a bit so that you can go for excursions if you so desire. Be warned though, some places are so interesting you may actually want to work longer hours. It is always worth asking a host in initial communications what is expected of wwoofers with regard to hours. Be prepared for a whole host of answers!

Building Work

Whilst building shelter is not essentially a key component in organic farming, it is not uncommon for hosts to be interested in ecologically sound building methods. Therefore, you may get involved in interesting eco-build projects during your stay. This can be a great deal of fun and another wonderful opportunity to learn about providing your own shelter. You may be preparing a dwelling for future wwoofers or perhaps helping to build a place for the hosts.

Types of dwelling you could work with or stay in may include Teepees, Yurts, Tree houses, Straw bale houses, Round houses, Cob and many other beautiful abodes. If you are not interested in working on these kind of projects, it is another good reason for checking first, before you go, what sort of work needs doing.

*There is space between the pages
To record your thoughts for ages.*

CHAPTER 6

SAFETY

Most hosts working practices are sound and it is very unlikely that you will sustain any serious injury whilst wwoofing (that is a relief – eh!) It is even less likely if you bare in mind the following.

If you are wwoofing for a while, at some point you maybe asked to undertake a task that you are not comfortable with or feel is dangerous. This may just be down to your own anxieties or it could be that the job is not actually a safe one. Having insurance may help you with treatment or travelling home, but it will not actually take away the pain of a thoroughly inconvenient accident. So if you do not feel comfortable with a task try to assess whether or not you could do it under guidance or if you would rather not do it at all. Hosts are usually incredibly accommodating and are likely to respect your decision. Do not be too hard on yourself. It may take some time to toughen up before you are able to ride wild animals unassisted! Taking out an insurance policy that *will* actually cover you for the type of activities you will be engaging in is still worthwhile.

Try to be flexible. The whole 'Health and Safety' mania that is sweeping the world may not have actually got to your farmer. They rely heavily on common sense and the ability to think on their feet in practical ways. They offer you a unique opportunity to develop those skills and may well be still doing so themselves. People of the land often have to come up with solutions utilizing whatever materials they have available. A supplier may be miles away and the money even further. Having said that, it is far better to turn down a task than just say 'yes' to using a tool that you do not feel comfortable with. Your host may not know what you are capable of. Furthermore, neither may you! Take care.

All of the suggestions in this book are just that, 'suggestions' – so please do not say to your host " 'The Practical Guide to Wwoofing' says that I can turn down this particular task." You might be told to get lost! It is a *guide*, to help you on your way. Make you own choices and decisions and don't lose your head.

The Little Old Man

I was asked to go up a ladder and pick cherries. Some of the steps looked like they were about to fall off. I shook my head at the host, smiling – as I did not speak his language. He tried to usher me up and I gestured to him that he was half my size and perhaps it was okay for him, but not for me. He did not mind and promptly raced up. He planted the tree 60 years before and had been picking it on the same ladder ever since it had been fruiting.
As he worked, the ladder seemed to just rest in thin air. I had not yet acquired the skills that he had and was glad to turn down the task, rather than risk serious injury.
I thought about repairing his ladder for him, but realized that perhaps he would then fall off. As he had aged, so had his equipment. He was as much part of the ladder as he was his beautiful Hungarian farm.

There is space between the pages
To record your thoughts for ages.

CHAPTER 7

LIVESTOCK

Working with livestock is a skill developed over a lifetime. On nearly every farm that I have been on that kept animals, I have heard stories of accidents or fatalities (usually on other farms!) It serves as a reminder that no matter how well we think we know the beasts; it is always possible to be nibbled, crushed, maimed or killed. A short number of months working with livestock can help develop a firm body confidence that naturally transmits to the animals. They are usually not interested in arguing and are more alarmed if *you* come across as frightened. If you are, you are unpredictable and intimidating. Their anxiety level rises and *they* are likely to become unpredictable. Personally I am not an expert in the field, though I have noticed that chickens, geese, pigs, horses and cows (amongst most others) are usually thinking about one thing when you arrive – FOOD. Can you be food to them or are you the bringer of food? Or are you going to make them into food?

As you check your safety, they are also checking theirs – to see if you are a threat to them. Survival is their priority. Humans sometimes forget this basic need and get injured trying to stroke the nice fluffy things. Even when doing this through a gate, a large animal can still lift its head suddenly and snap your wrist.

If you are being pursued by a large beast, (other than the tax man/IRS or the government) natural preservation will often cause them to run from you if you chase them – especially if you appear to be bigger than them (this does not usually work with the tax man/IRS or the government). I do not suggest you try this, unless you are absolutely confident in doing so – and need to.

An animal will sense everything. You must know what you are doing. Sometimes, just spreading the arms aside can give the impression of grandness and will ward off attacking geese or overly curious cattle. If in doubt – get out! It maybe too late once you are on the ground being trampled on.

During your experiences on farms, you may get lots of opportunities to learn more about beautiful creatures of the land. Keep calm, follow good guidance. Enjoy the splendour of such wonderful teachers – animals.

There is space between the pages
To record your thoughts for ages.

CHAPTER 8

ACCOMMODATION

What you stay in can be one of the most exciting aspects of wwoofing. An annexe, above a river. A tepee on the side of a mountain or possibly even a tree house. Often wwoofers stay in a caravan, sometimes placed in the most idyllic settings - and sometimes placed in the most dire. Most accommodation is basic, but occasionally you will be offered a very comfortable room in a farmhouse, really getting to sense the daily routine of the farmers. Regardless of whether or not you are away from the house outside or in with the family, bathroom facilities are often shared. However, if you are staying far away from the house, you will sometimes have your own compost toilet and in warmer climates, an outside shower. Or perhaps you will have the joy of just having the soil and the rain as your facilities! When you part company with your hosts, leave your room in at least as good a condition as you found it in. If it can be ready for the next wwoofer, all the better. One less job for your host.

CHAPTER 9

MEALS, FOOD AND DIET

Sometimes you will enjoy the luxury of having your food prepared for you. Other times you maybe provided with the ingredients and asked to rustle up your own meal. One of the most common food set ups is to communally prepare meals with other members of the family, wwoofers or workers. This makes light work of the task and is also a great time to chat.

Eating together is also an activity which does not always happen in many fast and frantic lives – even in families, for many reasons. However, whilst wwoofing I frequently found myself around a table with many people enjoying a meal that we had grown together, prepared and cooked and finally eaten together. Often with a home made bottle of wine. What joy to share such a time and space.

It is worth noting that just because you are wwoofing, it does not mean that you will always be given organic food to eat. In most cases meals are exceptionally fresh and local, often organic – though sometimes you may be eating supermarket packet food. For example, if a host's business is making jumpers from organic wool, they may live on a farm and keep sheep. The wool may be organic, but growing organic vegetables may take secondary place in their lives. Perhaps even further down the list or not at all. If you are really interested in what you are eating or perhaps have special dietary requirements, it is essential to speak to your hosts before you go, to make sure they can cater for you.

If you do not speak about dietary requirements before you go, your host may well ask you on arrival if there is anything you do not eat. It is fine to mention exceptions then but it maybe a little late to tell them that you are vegan (unless by chance they are too).

If you are vegetarian and other people consuming meat bothers you, think twice before you arrange a stay on a farm that keeps chickens and pigs – they may well eat them.

It is always handy to be able to answer the question "Is there anything you don't eat?" with "No, I eat anything", but do not worry if you cannot, hosts are often remarkably hospitable and may well have had many wwoofers before you- perhaps with special dietary requirements.

Handmade

Most of us routinely buy our daily provisions from a shop. Life changes when we make our own. Our own bread and cheese, yogurt and jams. We step into a different rhythm – perhaps a more natural one. When you wwoof you are likely to come cross at least a few handmade items. Whether it be a piece of furniture, food or a whole garden. Perhaps one may even feel the love that has gone into making a piece of work, the hours of 'patching' as you lay snug under a handmade quilt.

I love remembering the many 'handmade experiences' that I recall from my travels. The simplicity of such lifestyles, the peace that oozes from such recollection. Though to actually experience the bliss of 'hand craft' can reveal the primitive beat that runs through us all. Prepare yourself for a feast unto the eyes, in fact to all the senses when you enter the world of small holdings, farms and communities. Wwoofing is a window through which we can glimpse another kind of world.

CHAPTER 10

THE SETTING

Before you choose your wwoof hosts it is worth considering what kind of setting you would like to work in. If you are predominantly a town or city dweller, then being in a remote area without electricity or phone connection, may come as a bit of a shock. Or perhaps this type of setting is precisely what you need. Likewise, a farm next to or near to a main road may become irritating if you are looking for a peaceful place. Have a look on a map to assess these things before you arrange a stay. 'Google Earth' will give you an idea of urban or rural density, if you cannot lay your hands on a traditional paper map.

The size of the land a host will occupy varies considerably. From a large back garden in a town, to 50,000 acres of practically wilderness. The space may be occupied by one lone land keeper or hundreds of people in a community. Are you needing real solitude and quietness? Hustle and bustle - buzzy activity?

Or a family atmosphere? Children around? Other wwoofers? Sometimes it is difficult to know if you have not wwoofed before. Once you get started you may begin to form some ideas. Try to begin by imagining the sort of place you would like to be at. Be realistic, yet know that out of the multitude of wwoof venues there are around the world, it is very likely that there will be many that are right for you. It is not uncommon for this to be right on your doorstep.

Settling In

On the rare occasion that you feel uncomfortable at a place, first try and work through it. If you are really disturbed by an incident or the environment that you find yourself in, remember that you do not have to stay there.

Talk to your hosts if you can, but at least tell them if you decide to leave. There are sometimes a few words in front of your host list book regarding contacting wwoof in that country, if you wish to let them know about a particular experience that you have had, good or bad.

Likewise, a host can ask you to leave if your behaviour is unacceptable to them. This is rare. As with any trade or employment, some workers are not suited to it and neither are some employers. You get good ones and bad ones. In my experience, stunning settings with really lovely hosts and very willing wwoofers, far outweigh any unsuitability in this incredible movement. Considering that it basically works on trust and goodwill, you will soon experience yourself the wonder of the wwoof world.

There is space between the pages
To record your thoughts for ages.

CHAPTER 11

MONEY

If you do stay at one place for a while, there maybe opportunities to do other part-time jobs outside of the small holding, farm or centre where you are wwoofing. This gives you a chance to earn extra money. It may also give you a sense of independence and reduces the dependence on the host, giving space in the relationship. If you are wwoofing in your own country, it has been known for wwoofers to register as unemployed, actively seeking paid work and receive financial benefit from the government. In some countries it is possible to be a volunteer and receive this - something you may research.

French Alps

I was wwoofing in the foothills of the French Alps and had enjoyed such pleasure for the previous 10 weeks. Other wwoofers came and went, I moved from neighbour to neighbour depending on where work was most needed to be done. In my spare time I helped other local people saw up wood and earned some extra euros for the tasks.

We were far from internet cafes, phones, shops and bars. This had a strong effect on the pennies I kept in my purse, for they decreased little. So when my time was done in this snowy tiny haven....I counted my money to see where I could go to next and to my pleasant surprise I had spent during this two and a half month period a total of £40 pounds. I remember those days some years ago fondly now, particularly as it is now easy to spend £40 and some in just one day. Wwoofing may help us learn the value of money, often seeing much time pass with little actually passing through the hands. Other methods of trade take place, in more traditional and ancient ways. Particularly that of exchange or more often than not – just simply not buying that which we do not truly need.

There is space between the pages
To record your thoughts for ages.

CHAPTER 12

LANGUAGE

It is possible that your host will speak the same language as you and extremely likely if you are wwoofing in your own country! However you may often find yourself in difficult situations if you are not. Many hosts will make a real effort to communicate with you, but you can also meet them halfway. If you are travelling to a country foreign to you, it is worth trying to get to grips with the basics of the language. It can be very uncomfortable sitting around a dinner table whilst a family are talking away and you do not understand a thing. So if you are naturally inclined to the art of languages, try and pick up at least the basic courtesies before you go. Above all, do not worry. Man has wandered for thousands of years and found ways of understanding and being understood.

There are many other ways to communicate other than the spoken word. Drawing or even acting out what you are trying to say can be very effective. With regard to the help you will give on the farm, it is important to try and understand instructions given to you. Do not be afraid to ask the host to explain them to you. This *can* lead to the host just repeating the same foreign words again. However, if you follow the actions they are making involving tools or materials, you are very likely to succeed in carrying out the task successfully. And soon you may build a rapport with your co-workers, easing tensions around language.

Wwoofing is an excellent way to learn or develop language skills. The biggest step is the first, deciding to go overseas. The second is choosing the subject of your work – 'wwoofing', by now you may well have taken that decision too. Being comfortable with communicating will come in time. Do not panic – or if you do…put your tools down first. The first edition of this book had more mistakes in it than it had correct passages, it did not matter. It was written in 'pigeon English', a language that may be understood by many – perhaps as a result of being amongst people whose native tongue was not English. The point is, our intention was to communicate – and we did. I find it more comfortable to try, even if my grammar is wrong and be understood, than I do to be 'correct' but not understood.

*There is space between the pages
To record your thoughts for ages.*

CHAPTER 13

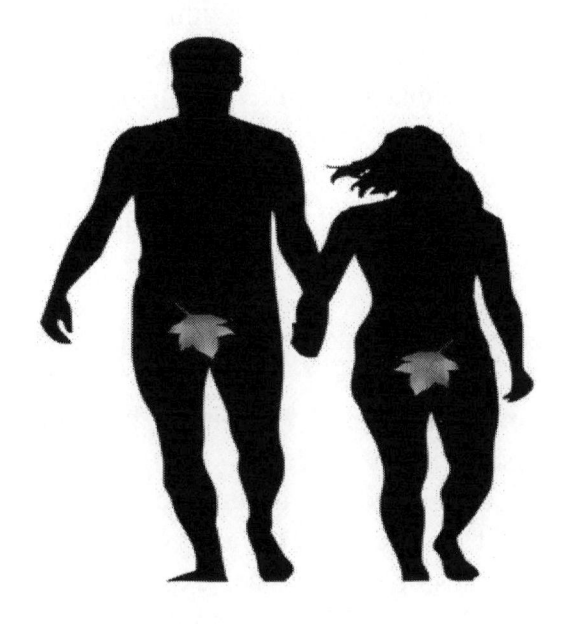

SEX 'N'DRUGS'N'ROCK'N'ROLL

The legend in the guide may indicate the host's policy on the use of alcohol - and to in some cases tobacco and drugs. (Sometimes it will not be mentioned at all) It may fall somewhere on the scale of: "accepted, tolerated, not allowed". If hosts have strong views on the use of these, they will usually tell you before you arrive. On the other hand, you may discover only on arrival that *they* smoke tobacco or cannabis (the latter being considerably less common). Perhaps even indoors, so if you feel strongly about being around these you could mention before you go that you do not smoke. This may bring up the subject before you arrange your stay.

Alcohol is more common, but has yet to have been a problem on placements I have experienced (with the exception of once when it ran out). I have not seen any excessive use of it anywhere.

Most wwoofers are young and single but also extremely transient, so if you do hope for a bit of 'up and close', go in with your eyes open if you are expecting a longer term relationship. One of you may soon move on, perhaps even back to your own country. Having said that, I have come across quite a few couples who met wwoofing and travelled on together. Often then settling in one place and raising a family. Then perhaps even going on to receive wwoofers themselves. If you are lucky enough to wwoof at a place like this you may benefit from volunteering for someone who has seen both sides of the coin – to be a wwoofer and a wwoof host.

Of Love; Wwoofing in Spain

I met a girl from Hungary. She invited me back to her country to wwoof and I accepted this exciting opportunity. The cultural differences were immense. Fortunately, the organic farming had much in common with many other places that I had been and so it was the land that held me together. The girl soon returned to her home - the city and I struggled to manage in the environment. I followed for a while but could not survive in such a place. The love soon fell apart. I carried the pieces of my broken heart back to England, where I wwoofed around 15 farms. It was a time of healing and much stitching back together of wounds. Working on the land, clearing and sowing brought about a sense of wholeness and clarity once again. The next time I met a woman on a farm – I walked the path with my eyes open (as we had met by bumping into each other when my eyes had been closed!)

There is space between the pages
To record your thoughts for ages.

CHAPTER 14

ARRANGING YOUR STAY

There are 4 main possible ways to arrange your stay or 5 if you include carrier pigeons, or 6 if you send a message with another wwoofer..etc. By phone, email, letter or personal visit (by arrangement). Some hosts may not have access to internet and some may not have a phone.
It can be a bit challenging to get your head around this if you are deeply submerged in the modern world and flooded by technology yourself. Try and imagine a place that does not even have electricity! Not even solar or any other form of alternative power sources, other than nature of course. Being at a wwoof venue with such simplicity can be excellent for health!

If you do live close to a possible host, it is worth a visit to see if you might fit in there. Remember that a host may receive many enquiries and may take time to reply. However, if you do not hear from them within a few weeks, it is okay to follow up with a phone call. The host's description may also indicate their preferred method of contact.
You may need to give a few months notice for a placement. All stays are arranged directly with the host by you. In some cases, a host may only need a week or even a few days notice by phone, especially if you sound really handy and they are in need of help at that particular time. Make sure you mention any relevant special skills that you have and perhaps areas that particularly interest you about their farm or subjects that you would like to learn more about.

There is not usually an official interview before volunteering (unless otherwise stipulated). Both parties try to gauge whether or not the partnership will be a good one. So in a way, the contact itself is where the assessment is done, both sides looking at the viability of the proposal. If you like the look of a description, do not necessarily jump straight in – make some enquiries first. If the response to your enquiry does not seem that positive or encouraging – maybe find another one. My own experience is that if a place does not feel right for me, it probably isn't. If you are able to trust your intuition, it is a good opportunity to do so. Keep going until you find a place for you. Remember that a host's description may well have been written a few years ago and although ideally it would have been updated, perhaps it has not been. Therefore, try to be specific if there are particular aspects of the placement that interest you, just incase that part does not feature anymore and was the reason for your proposed visit.

There is space between the pages
To record your thoughts for ages.

CHAPTER 15

TRAVELLING TO YOUR HOSTS

Buses and trains.

Whilst a 24 hour train journey will cost the equivalent of £1 in some countries, in others a 2.4 mile journey may cost the same. It is worth spending a little time researching prices before you head off on your wwoofing journey. The connections from host to host can really add up. Of course there is always the possibility of hitchhiking – we will come to this later. Booking a bus or train 30 or 60 days in advance, via the internet, may lead to substantial discounts. A journey from Scotland to London by train was set to cost me £150. Buying it one month in advance online cost £19. Some disadvantages of buying in advance maybe that you lose some flexibility to take up spontaneous opportunities. Perhaps offers and invitations to stay with other people that you meet or take up paid work, for example. However, with low cost tickets you have not lost too much if you do not choose to use them.

By Boat

Going as a foot passenger on a ferry can be a very cost effective way of travelling – assuming there is a sea to do so. Costs can mount up if you need to stay in a cabin or buy food aboard. If you are of limited funds and the journey is a short one, then pack some food and find a comfortable seat – reducing the costs substantially. Travelling by boat is worth considering in certain cases, though is one that is often overlooked. Boarding a ship can be far more relaxing than going through airports and throwing your body through the air at 500mph. Not to mention what you are doing to the air that you breathe.

Purchasing a Vehicle

You may choose to buy the vehicle in your own country, if you can take it on a ferry or if you are staying put and wwoofing there. Or perhaps you will acquire one in the country you arrive in. The obvious advantages of having a car are your independence and the option to be able to pick up hitchhikers - if you feel safe to do. Also the amount of room you will have in which to carry equipment and food. It may also be possible to sleep in your car, especially if it is an estate. The costs involved in repairing and running a car can be quite high and for this reason it often works well with a couple. A small van can be even more practical. One of the downsides of owning a vehicle is that you will frequently pollute the environment and often the food that we all eat. However, it is quite a challenge to live in an environmentally sound way and in a practical manner, so nearly all forms of transport you use will have a strong implication on your environment. This is an extremely complex matter – not one we aim to conclude in this book. I have cycled in many countries, but even with this method I found that the effect of being hit by a car was far more disturbing than driving one. Perhaps your wwoofing journeys may help *you* to make a balance.

There is space between the pages
To record your thoughts for ages.

CHAPTER 16

CYCLING

Travelling by bike is a great way to see the countryside under your own steam and a free way to get from host to host. It also means that you may be able to carry a bit more equipment than you can on foot for example a tent, food and little cooker. If you are going to go this route, consider some of the following:

Choose a bike that is comfortable for you, not just a mountain bike because it looks cool. If it needs to be a more upright road bike then so be it…or if it needs to be a racing bike, then so be that too! Perhaps hunt around for a good second hand bargain before you go, rather than spend half your journey money on a new one;

Make sure you know how to do at least the basics in repairs. Change a tyre, fix a puncture and replace a snapped cable;

If you can tune up gears, replace broken spokes or straighten a buckled wheel – all the better. You may find a bike shop on your journey – though this could render your main mode of transport out of action until there is a slot to repair it. Or you maybe stuck in the middle of nowhere and not be able to get to a repair site, so brush up on all the skills you can. On the other hand you maybe a bit of a 'bike wizard' already. In which case you have a valuable trading commodity. Bikes often need attention and people have them the world over. This may help fund your wwoofing journey or at least be helpful to wwoof hosts who have bicycles lying around that need attention.

CHAPTER 17

THE JOURNEY – ON FOOT

Naturally it is possible to walk from host to host or to a hostel or campsite. Or perhaps you will sleep out in the countryside (if it is safe.) If you do this make sure you are well equipped. Not only materially, but also with the right knowledge to look after your basic human needs. Try not to upset local people by camping or building shelter where you are not wanted. If you are doing this in a country not known to you, find out if there are any natural predators first. Perhaps the greatest threat posed to man is man himself, though poisonous insects and snakes can also be a little irritating. If you are not camping in a tent, what kind of shelter can you make? Can you take it down with ease? Leave areas that you stay in at least as clean as you found them. Are you able to keep off the ground? Perhaps carry a simple hammock.

If you are in doubt about 'sleeping out', try it first in warmer weather. If you are still in doubt, do not do it. Do you know how to make a fire, safely? Does it matter if you draw attention to yourself with it? There are many other points to consider when travelling on foot. For example the amount you can carry comfortably. How much food and water can you take with you and where is the next source of it?

All of these factors can be dealt with but make sure you know how before you set out on your epic journey on foot. People have walked the earth for an age, some get sick or meet fatal ends. Others have an incredible journey and never regret it. Perhaps strike a balance between the nomad's way and the practical path. Perhaps not. Do it your way. Go well.

<u>Greek Islands - Under the shade of an olive tree.</u>

It dawned on me that dusk would soon be falling. The sizzling olive groves would keep a steady comfortable heat through the cool night. A few miles earlier, a local man had suggested that I could find a place to sleep in between the trees up the road. I placed my tent upon the warm ground of this Greek island. In the morning I unzipped the front, only to see a man walking past with a machine gun. He turned to me, smiled and walked on. I had slept on a shooting range. Humour and perception may vary from country to country.

There is space between the pages
To record your thoughts for ages.

CHAPTER 18

HITCHHIKING

Most hosts will pick you up by car from the nearest bus or train station. Though some may not own a car or you may not have the money for public transport. If you are up for a little more adventure, consider hitchhiking. In order to do this, you must feel confident and safe with the idea. Do not try it otherwise. It is not for everybody. Making your own way to a farm can be rewarding, satisfying. Likewise, finding your own way out when it is time to leave. Be warned though, hitchhiking can be an exhausting method of travelling, taking many rides to get to your destination. Often waiting hours for a lift and of course hiking with your pack. On the up side it can be really interesting to meet completely random people who at that time respond to your request and give you a space in their lives for a short time.

I have had the most wonderful conversations with drivers and only occasionally had an undesirable ride.

A few points to consider:-

Carry a mobile phone if possible, food and water;

Be safe, do not take risks here. Start early – giving yourself plenty of time to get a ride;

Stand in a position where the car can easily stop;

Carry a map, know where you are going;

Write a sign if you feel it will work e.g. 'London';

Have a back up plan – it is no fun sleeping at the roadside;
Let your host know that you are travelling in this way;

A man and woman hitchhiking together are often more successful than a single man;

Know that you can turn down the offer of a ride;

Thank the driver for the ride!

*There is space between the pages
To record your thoughts for ages.*

CHAPTER 19

BAGGAGE

If you are only going to one host, the pack you take with you need not be too big. Mainly because you will have a good idea about the season and the type of clothes to take with you. However, if you intend to go from one farm to the next for a greater length of time, there are many factors to consider. It is possible to buy suitable clothes if you change countries or move through different climates. Or even to post things back home when you do not need them. Preparing the right pack is an art, one you will appreciate and develop as your journey goes on.

One reason for taking a lot of items may be the comfort we derive from having music, books and even computers with us. The list goes on, extensive amounts of toiletries spare clothes, writing sets …

It can be quite a task to get from place to place if we are laden down with a heavy pack. Take courage and minimize. You will be amazed with what you can do without in order to just meet your basic human needs. Instead of asking your self "What do I need to take with me?" ask "What can I do without?" The physical gaps left by things you have not brought with you will soon be filled by spaces of experience that are on offer in your new daily life by the host. For example, when live music comes as wwoofers gather, it is a treat to hear – rather than carrying with us the endless amounts of electronic entertainment gadgets that we may be accustomed to.

Again, when we are offered internet time – you get down to it and just send concise messages, perhaps only once every few weeks instead of your usual use, which might be considerably higher. In these ways we can make the whole wwoof/travel concept a lighter one. Resist the fight and travel light! These days it can be very useful to check on a mobile device with internet access on it, to confirm travel arrangements or your next wwoof host confirmation. The point is that we need not be dependent on it exclusively in order to occupy our time. You will know if you start to make headway with these obsessions when you find yourself looking at an animals face, in real life, instead of on *Facebook*. Hooray!

The Pack Itself

Avoid the really cheap bags that do not have much padding on the straps. A back pack is the most popular choice. There is no need to have a really expensive one, it is more likely to get stolen, usually in transit. Clean second hand ones are not that common, but are obtainable if you are on a budget or just want to recycle something and use one that a friend or family member has just sitting in the cupboard. Or check local newspapers and the internet for adverts in your area. I was lucky enough to find my latest one in a second hand shop for just a few pounds. My previous bag was given to me by someone who gave me a ride in New Zealand. Layout your contents and choose the right size bag for you. You will live out of it for the coming weeks, months or years.

There is space between the pages
To record your thoughts for ages.

CHAPTER 20

EQUIPMENT NEEDED

Lessons in less.

When I first started wwoofing I carried with me a pair of Wellington Boots and a comfortable indoor shoe or casual going out 'trainer'. I also had a pair of good walking boots. As years passed, I got tired of carrying them all around with me and wanted something that would do the lot – footwear that may cover all eventualities.

I thought about how useful 'wellies' were but also what a task it was to get them out of half a metre of mud whilst bulls or tusked boar were nudging me, asking for food and wondering if I could become it. After all, my feet could breathe at night time, bare in the air and I could also wear a thick pair of socks indoors for comfort, instead of having slippers! Eventually I opted for an army boot, which I dubbed most frequently to waterproof. As I travelled I used anything I could to keep them watertight, lubricated and soft. Vaseline, olive oil or saddle oils – it mattered not. If there was nothing about, as a last resort, I would actually buy some real dubbing! I was not working in any hot climates at the time so they did not become uncomfortable. In such places, even a water proof sturdy walking boot would do. (Some people really have a good relationship with 'wellies', if you are one of them then of course stay with your rubber beauties!)

Soon the idea of travelling with little baggage spurred me on to think about what else I could do without. Could I manage with only two sets of clothes, one set to wear whilst I was washing the other? Yes. It worked very well and my pack got smaller and smaller. Eventually I was able to walk miles with it. I also carried a good tough old leather jacket with me that would protect me from rough wwoofing work and nibbling livestock and was able to oil and waterproof that too. The army surplus stores where I bought my boots from also sold high quality low priced foul weather trousers and jacket too, so for severely wet skies I was well covered.

The climate you work in may well determine much when choosing equipment or clothes. The mindset you go in, even more. Hosts normally have a selection of boots, jackets and other items, but if you are serious about working the land, it is worth carrying your own basic kit that you are not only familiar with, but also really comfortable with. The following are well worth investing in:-

EQUIPMENT NEEDED

A small lightweight torch either halogen bulbs or wind up, to save on batteries:

A tough pair of gloves, high quality leather. Close fitting gloves will do for weeding and keeping warm;

A good folding knife. Great for cutting vegetables and small wood. I recommend the 'Opinel No.9', but there are many other good ones to available;

A small sharpening stone. If you know how to sharpen your knife, it will always be ready when you need it.

A map of the area you will wwoof in. Small 'A-Z' map books of an entire country can be an efficient type to use;

A needle and thread. Wonderful if you know how to sew. Small holes just get bigger and let more things inside your clothes!

A sleeping bag. Try and get a natural fibred bag, artificial ones can cause you to overheat;

A sun hat. In hot climates, it is a must. An afternoon in strong sunlight can bring on severe sunstroke. In rain a well made one is also useful;

Hosts often provide bedding. I have rarely needed to use a 'sleeping roll' or carry one in my pack whilst wwoofing. The exception is if you are planning on some independent camping in between hosts. Then you will need one, unless you can make your own mattress. There are other things you may need. This section serves as a key indication only.

India

In between wwoofing, one project I embarked upon was to head over to India and buy an old fishing boat. I knew that it was impractical to take tools or equipment with me and that I could buy all those things over there. I wondered what else I could do without taking. Clothes perhaps? (Apart from the ones I was wearing). Almost everything I could buy in India would be considerably cheaper than it would be in England, so I set myself the challenge of leaving home with the smallest bag possible. I ended up taking a bag no bigger than a piece of A4 paper, (well perhaps the size of an A4 pad) containing a toothbrush, medicine, razor, one pair of pants and socks, a guide book and money. On hindsight, I could have done without the guide book and razor. I bought 2 fishing boats, nets, timber for modifying them and of course more clothes, for a total of £200. But aside from being able to buy cheaply in many countries, some travels may simply not require much luggage.

It is a liberating feeling to just travel with a day pack, though it can take an entirely different mindset to do so. It is not suitable in all situations and may actually land you in a great deal of trouble if you have misjudged the needs of your trip and your ability to acquire what you need en – route. However, where it is appropriate (particularly in between wwoofing) this light experience can be extremely exhilarating.

There is space between the pages
To record your thoughts for ages.

CHAPTER 21

IN BETWEEN WWOOFING

Earlier on in the book we talked a little about leaving gaps and flexibility in your journey. In this section we shall delve much deeper into the subject. Being open to possibilities may mean the extension of your journey and for many people this is an exciting prospect. If you are wwoofing for long periods of time it may also be attractive to break up your journey a little with other types of work or activities.

If your face fits, once you have been in an area for a while, it is not uncommon for people to ask you if you would like to 'housesit' or even landsit (look after a house, a piece of land or both). This is particularly likely to happen in countries where people have purchased second homes, small holdings or parcels of land as they may not always be there to watch over them. It may also be an opportunity to save money in between wwoofing, whilst the costs of living during house sitting are at a minimum.

Spain

Whilst wwoofing in Spain I met an Englishman who owned a riverside ruin on a couple of acres of land. Oranges, lemons and almonds grew in abundance. As we got to know each other I learnt that he was working on the property about 6 months of the year. The rest of the time he would return to the U.K. and needed a 'land sitter' to watch over the place. There was fresh running water on the site, a few walls to the house and very little else – but what else should I need in such a glorious Spanish summer? Very little.

I took on the property and set about putting in a vegetable garden and made the ruin good for staying in. Cooking outside on a barbecue fuelled by fallen twigs from the trees or pruning's. I had been wwoofing for quite some time so taking on this responsibility gave me a sense of independence. I made some furniture from 'found' timber and managed to sell some too. I had practiced T'ai Chi Chuan quite extensively (a gentle Chinese moving meditation –originally a martial art) and was able to give some lessons for a few Euros. In this way, my journey went on and I had some time off from wwoofing whilst saving some money for another trip.

Hungary

Whilst wwoofing in Hungary, I taught Conversational English to some local people. They were keen to pay to improve their spoken word, though initially I kept my prices quite low. I did not have a teaching qualification and my grasp of grammar was poor. However I was friendly, spoke English quite clearly and had taught before in Latvia and Brazil.

After a time I was offered some quite highly paid work to brush up the fluency of business peoples English in the city. Though cities are not my natural habitat, I realized that the work would enable me to fund other voluntary experiences where I may stand to learn new skills.

In itself, it was also an interesting experience to suddenly be in plush offices and zipping around manic public transport systems. It reminded me of how important the countryside was to me and how much fast moving things and thick black air were not.

<u>Poland and Latvia</u>

In a hostel in Warsaw, Poland, I noticed one woman who reminded me of an actress, perhaps 'Julie Andrews'. I got chatting to the lady, discovering that she represented an orphanage in Latvia and she offered me a job as a volunteer. Within 3 days I was there.

The real tough years at the home had passed. The young kids were now teenagers. They had good clothes, schooling and even pocket money. Naturally, nothing could replace their parents but they were much happier than of previous years. We played outside frequently and although it was awfully cold, we had a lot of fun, especially building igloos.

The building was huge and impressive and had been lovingly restored in recent years, with a wood burner the size of a steam engine to keep it warm...and so it was. Outside, it was still -17 degrees! Originally they endured cold indoor temperatures too.

One of my indoor jobs was to speak English with the children. Another was to help restore old furniture, stashed away in an old barn. I volunteered for a few weeks in exchange for a great room and tasty food. Lovely Latvia, crisp and bright.

What you do before, in between or after wwoofing may be just as important as wwoofing itself...or more so. In some respects it may show that you can be self-sufficient, or at least are willing to pursue what interests you most, even if you are not entirely successful in your ventures.

Beginning a Journey.

As the last millennium drew to a close, I planned a journey to a far away place –to the dusty sunlit mountains of central Brazil and soon set off on an adventure that would completely change the direction of my life. I ambled around for a time before winding up on eastern shores, where sea met jungle and I met joy. I needed a place to lay my head for a while and opted for the most exciting option – a tree, a place where I could build a little shelter whilst having stacks of fun.

So it was to be that I strolled some further miles in search of the perfect 'house holder' and before long met a man who had such a tree – fit for the most splendid tree house.

Driftwood lay in abundance on the hot white beaches and before long the little jungle edge community lent a hand in knocking up this simple shelter. Weeks passed and whilst the tree home took shape, I slept in a tent underneath this perfect tree, amidst sandy roots and giant ants.

And when the time did come to climb the ropey ladder and rest my sunburnt head, I crawled up into the little dwelling and was gently rocked to sleep by the wind teased branches. Days were spent swimming in the surf or canoeing through creeping mangroves. Then when I tired of this, I split the whites of coconuts with my proudly held shimmering machete.

Until the time came when the heart of England called me. Return I did, if only to see that I did not want to be there. So it came to pass that I cycled my way down the interesting land of France, where my wwoofing journeys began.

CHAPTER 22

JOINING WWOOF

Contacting a Country

So you have chosen which country you would like to wwoof in. Now look at the list that follows. Find that country in the list. Do they have their own wwoof organization? Contact them to see what they need in order for you to join. It is usually a completed application form, a photo and payment (£10 to £30). They will explain to you how to do this. They will then send you your membership details along with a list of hosts who receive volunteers. This will arrive by email or by post.
Some countries do not actually have their own wwoof organization that you can join. These are all clubbed together in one group, called 'independents'. One of the website addresses where you can join independent countries appears frequently in the list that follows. It is:

www.wwoof.org/wwind/

Another good website to use where you can also buy an 'independents list' is:

www.wwoofinternational.org/independents

They have a great interactive map, which is really helpful.

They both offer similar membership and access to the hosts contact details that do not have their own wwoof organization. Use whichever website you feel most comfortable with. Increasingly countries that do have their own organization are also giving the independent list when you join, for example Italy.

So it really is worth having a thorough look around and researching all the places that you are interested in before you jump in. Or jump in…get going and work it out as you go along!

At first it can be confusing to understand all of these options as presently there is not one united WWOOF organization. However, once you start to make enquiries things may become a little clearer. If not, it will come organically. The following list, although a little repetitive – makes it clear whether or not a country has a national organization and who you should contact.

Country Contact Details:

America (Latin)

www.wwooflatinamerica.com

Argentina

www.wwoofargentina.com

Australia

www.wwoof.com.au

Austria

www.wwoof.at

Bahamas
www.wwoof.org/wwind/

Bangladesh
www.wwoofbangladesh.org

Barbuda
www.wwoof.org/wwind/

Belgium
www.wwoof.org/wwind/

Belize
www.wwooflatinamerica.com

Benin
www.wwoof.org/wwind/

Bolivia
www.wwoof.org/wwind/

Brazil
www.wwoofbrazil.com

British Virgin Islands

www.wwoof.org/wwind/

Bulgaria
www.wwoofbulgaria.org

Cambodia
www.wwoof.org/wwind/

Cameroon
Peter Njome.
Telephone: 233 244 773971

Canada
www.wwoof.ca

Croatia
www.wwoof.org/wwind/

Chile
www.wwoofchile.cl

China
www.wwoofchina.org

Columbia
www.wwoof.org/wwind/

Cook Islands

www.wwoof.org/wwind/

Costa Rica

www.wwoofcostarica.com

Czech Republic

www.wwoof.cz

Denmark

www.wwoof.dk

Dominica (commonwealth)

www.wwoof.org/wwind/

Ecuador

www.wwooflatinamerica.com

Egypt

www.wwoof.org/wwind/

El Salvador

www.wwoof.org/wwind/

Estonia

www.wwoof.ee

Ethiopia
www.wwoof.org/wwind/

Finland
www.wwoof.org/wwind/

Fiji
www.wwoof.org/wwind/

France
www.wwoof.fr

French Polynesia
www.wwoof.org/wwind/

Gambia
www.wwoof.org/wwind/

Georgia
www.wwoof.org/wwind/

Germany
www.wwoof.de

Ghana

Email: kingzeeh@yahoo.co.uk
Contact: Kenneth Nortey-Mensah
PO BOX 154, Trade Fair Centre,
Accra, Ghana
Telephone: 00233 244 773971

Greece
www.wwoof.org/wwind/

Guatemala
www.wwooflatinamerica.com

Hawaii
www.wwoofhawaii.org
www.wwoofusa.org

Haiti
www.wwoof.org/wwind/

Holland
www.wwoof.org/wwind/

Honduras
www.wwoof.org/wwind/

Hungary
www.wwoof.hu

Iceland
www.wwoof.org/wwind/

India
www.wwoofindia.org

Indonesia
www.wwoof.org/wwind/

Ireland
www.wwoof.ie

Israel
www.wwoof.co.il

Italy
www.wwoof.it

Jamaica
www.wwoof.org/wwind/

Jordan
www.wwoof.org/wwind/

Kenya
www.wwoof.org/wwind/

Laos
www.wwoof.org/wwind/

Latvia
www.wwoof.org/wwind/

Japan
www.wwoofjapan.com

Jordan
www.wwoof.org/wwind/

Kazakhstan
www.kazakhstanwwoof.narod.ru

Kenya
www.wwoof.org/wwind/

Korea
www.wwoofkorea.co.kr

Lebanon
www.wwoof.org/wwind/

Lithuania
www.lithuania.lt

Luxembourg
www.wwoof.org/wwind/

Macedonia
www.wwoof.org/wwind/

Madagascar
www.wwoof.org/wwind/

Malaysia
www.wwoof.org/wwind/

Mali
www.wwoof.org/wwind/

Mongolia
www.wwoof.org/wwind/

Mexico
www.wwoofmexico.com

Moldova

www.moldova.org

Morocco
www.wwoof.org/wwind/

Mozambique
www.wwoof.org/wwind/

Namibia
www.wwoof.org/wwind/

Nepal
www.wwoofnepal.org

Netherlands
www.wwoof.org/wwind/

Nicaragua
www.wwoof.org/wwind/

Nigeria
www.wwoof.org/wwind/

Norway
www.wwoof.org/wwind/

New Zealand
www.wwoof.co.nz

Palestine
www.wwoof.org/wwind/

Panama
www.wwoof.org/wwind/

Peru
www.wwoof.org/wwind/

Philippines
www.wwoof.ph

Poland
www.wwoofpoland.wackwall.com

Portugal
www.wwoofportugal.org

Romania
www.wwoof.ro

Russia

www.wwoof.org/wwind/

Senegal
www.wwoof.org/wwind/

Serbia
www.wwoof.org/wwind/

Sierra Leone
www.wwoofsl.org

Slovenia
www.wwoof-slovenia.info

South Africa
www.wwoof.org/wwind/

Spain
www.ruralvolunteers.org

Sri Lanka
www.wwoof.org/wwoofLK

Sweden
www.wwoof.se

Switzerland
www.zapfig.com/wwoof

Taiwan
www.wwooftaiwan.com

Thailand
www.wwoof.org/wwind/

Togo
www.wwoof.org/wwind/

Tonga
www.wwoof.org/wwind/

Trinidad and Tobago
www.wwoof.org/wwind/

Turkey
www.bugday.org/tatuta/?lang=EN

Uganda

Contact: Bob Kasule
Email: bob_kasule@yahoo.com
P.O. Box 2001, Kampala,

Uganda, East Africa
Telephone: 256346856, 251276

United Kingdom
www.wwoof.org.uk

United States of America
www.wwoofusa.org

Venezuela
www.wwoof.org/wwind/

Zambia
www.wwoof.org/wwind/

Wwoof Founding Body

www.wwoof.org

Useful Link: **www.helpx.net**

BASIC CHECKLIST

Check the safety of a country before you travel;

Join the wwoof organisation that covers that country;

Arrange your stay;

Check visa information before you go;

Arrange insurance and equipment;

Learn the basics of the language and customs;

Make sure you have enough money…or brains;

Make sure someone knows where you are;

Prepare for the adventure of your life!

There is space between the pages
To record your thoughts for ages.

CHAPTER 23

VISAS

Entering and leaving a country

It is very important to make sure that you have the correct visa for your wwoofing experience. This of course will not only depend on the country you have chosen to wwoof in, but also from which country your passport is from. We hope for a time where it matters not where you come from in order to volunteer on a farm, but until then – it may be worth considering two things:

The smooth transition through this stage may depend enormously on a factor that cannot be calculated beforehand and that is the frame of mind the official you meet at immigration is in on the day you present your papers to him or her. It has been known for immigration officials to ask the question:
"What is the nature of your business?"
And it has been known for wwoofers to reply with the answer: "I am a tourist", not mentioning wwoofing. A tourist may be asked where they are staying and it may be useful at this stage to have a reservation at a hostel. I make no judgement about this relaxed attitude and encourage you to stick to the law – wherever possible. Wwoof Australia had this to say about visas:

"In some countries over zealous customs officials have been known to put WWOOFers on the next plane home when told the reason for their visit is to go WWOOFing if they are on a tourist visa In Australia the Immigration Department is now fairly familiar with WWOOF, but a few years ago most customs officers didn't really know anything about it and were more inclined to send people

home. Their ruling for tourist visas is that WWOOFing is fine so long as it is not the MAIN reason for your visit, and so long as the work done would not otherwise be a paid position (e.g. Feeding the chooks is fine but picking a commercial crop is not)."Wwoof Australia

Ukraine to Turkey

One worries considerably about not being granted permission to enter a country. Also consider that you may not actually be able to leave. I was scheduled to leave the port of Odessa in the south of the Ukraine. One last visit to the toilet was called for before boarding the vessel that would take me across The Black Sea to Turkey. Unfortunately the public toilets were out of order and I wondered what course of action to take. Passing by was a distinguished looking gentleman wearing a naval uniform and I explained my predicament to him. Kindly he allowed me the use of the staff facilities, gave me the key and entrusted me to drop it back to the reception desk. As I attempted to leave the toilet, it came to my attention that the lock was jammed. For several minutes I tried to free it, but to no avail. I tried to squeeze under the bottom of the door...but no luck...I had been fed well by my last hosts. My boat would leave in a matter of minutes – what should I do? I called for help, nobody came. Finally, there was no other option apart from waiting...I had been beaten by a Ukrainian urinal, I must break the lock. I took one sharp kick at it, but it did not break.(The toilet design was as tough as the vodka drinking fellows I met on the boat).

Instead, it took the door off the hinges causing it to fall forward to the ground. Jumping forward I caught it. Oh gosh! Being an honest sort of chappy, I explained to the lady on reception what had happened. I don't think she really understood me as she just replied "Never mind". I ran to the boat to the sound of the last whistle call and regained some sort of calm in the quietness of my cabin...where I found vodka. By all means, make sure you can enter the country of your choice – likewise, make sure you can leave it.

CHAPTER 24

RETURNING HOME

Sometimes wwoofers do not return home, perhaps relocating to another country. (But do not worry mums and dads...they usually come eventually!).

Returning home can take some adjustment. Finding a place to stay, perhaps in an urban setting. This can be daunting after chunks of time in the countryside. No doubt your friends and family will be interested in your travels, though if they have not experienced anything quite like it, it could be a bit tough for them to relate to all of your experiences. There are many ways of dealing with this transition. It is often when people start looking at wwoof hosts nearest to home, reaching out to make local connections. Or maybe you could look at your next wwoofing trip, begin to plan it. How would you improve it? Different baggage? Walk, cycle or go with a friend? Besides wwoofing there may also be other volunteering work that you can do locally. Or perhaps you can start an allotment in your area or try and work a piece of common ground?

There are more and more opportunities around as many of us begin to realize the importance of eating good food that does not cost the earth. Perhaps you will contact your local newspaper. They may like to do an article about the wwoofing you have done. Or the 'Wwoof News' if you have wwoofed in the U.K. Some of the larger wwoof organizations also publish newsletters and hold meetings and gatherings that you can attend to keep in contact with the movement. Even if you do not pursue wwoofing, it can be another kind of work experience you may add to a C.v. which can be viewed as valuable life experience.

CHAPTER 25

MORE QUESTIONS AND ANSWERS

How did you come to write the book?

I was on a farm in the Lake District, England, thinking much about wwoofing when it just occurred to me that there may not actually be a simple paperback guide for new wwoofers. Over the coming months, I found lots of references to wwoofing in other books and also online information, but not a plain 'A-Z' type book. As I had wwoofed and worked myself around much of the world for the previous 10 years, I decided that I may make a contribution for future land liking travellers who wished to know how to develop skills in organic farming and self – sufficiency.

But there is loads of free information online about wwoofing.

Some people would still like to have an extensive ready made guide, either that they can carry with them as a paper back book or have access to a full document in as an e-version.

What did Wwoof think about you writing the book?

The wwoof organisations in each country were connected by name and aims and often operated autonomously. It seemed unnatural to contact the wwoof group in each country I had been to and ask them if I could write about my experiences and form an A-Z for new wwoofers.

So did you ever tell any Wwoof organisation that your book was out?

Yes, I contacted all of them and told them.

And what did they think about it?

Nearly every country that got back to me liked it, some giving me quotes that I could use about the book. Wwoof in the U.K invited me to their Annual General Meeting in October 2009, where upon I met **Sue Coppard**, the founder of Wwoof. I presented her with a copy of the book. Later she commented on it and said:

"I found it to be a delightful and informative read, very accessibly and humanly written – it almost reads itself!"

Why did you write the book?

I couldn't stop my self! Plus, in 2009 there was not another paperback available on the subject – a handy little book that could be passed around in a very traditional way. I feel passionately about what wwoofing can bring a person. Namely, that it can get us so much closer to experiences in natural settings and be excellent for well being. We also get an opportunity to learn so much about providing for our basic human needs simply and self sufficiently. As an exchange we can bring fresh energy to the place where we volunteer.

But these simple traditional skills are nearly lost, how does one go about learning them without paying a lot of money to do courses? Surely people on farms don't still live in such a simple manner?

I am sorry, either you have not read this book, or you simply don't believe the wonder of what is available to you until you experience it. It may be hard to believe that you can stay in a gorgeous old wooden cabin by a river that is clean enough to drink from. Feel the warmth of the sun on you back as you pick strawberries with a group of happy wwoofers singing as they help!

If you wwoof for any length of time, you are likely to be in close contact with valuable life skills such as:

How to grow your own food, provide fire to cook it and to warm yourself by. If you wwoof even longer, in a greater variety of places and countries you will probably gradually acquire skills such as natural building methods, so you could even build your own shelter. How to use natural medicine effectively and maintain a balanced diet for good health and much more. So it is possible to equip yourself with a wide range of useful skills whilst helping farmers as you learn, without paying a lot of money for courses. Naturally if you are not in a position to go off and volunteer on farms, courses are useful. Or perhaps do both. Consider though that Wwoof started out as: *'Weekend Workers On Organic Farms'* and that there are still some opportunities to go and help out for even a couple of days. Go and have a great time!

It seems that you are a bit of a wwoof enthusiast, do you really think that it is the miracle cure to save the planet?

Absolutely not, it may surprise you to read! I do not think there is just one way, single mindedness is more likely to lead to just irritating people. It is *a* way, that seems to be becoming increasingly popular. Not just wwoofing but returning to the land and working toward a natural, simple life. Perhaps it will need other ways too, in order that the planet will be saved. Maybe it will save itself, I think it is more than capable of showing us who is boss and perhaps we are just a tiny bit more insignificant than we think?

What's the point of doing all this 'pure stuff' when the planet is going to hell in a handbag anyway?

Well what point is there in it for you? For me, I prefer to breathe clean air, drink unpoisoned water, eat chemical free food and live in peaceful surroundings conducive to keeping a healthy mind, body spirit and soul – whilst trying to irritate as few people as possible.

Do you actually live a life that does absolutely no harm to yourself, another or the earth?

No. For example, this book was produced on a computer – the manufacturing process of such a device has astonishingly harmful effects to humans and the planet. However, in balance – reading the book may have a positive effect on some people. The paperback also has the huge benefit of only being printed on demand, for example when it is purchased it is then made. So it doesn't sit around in piles gathering dust. In this respect, it is on the greener side of publishing and naturally it is available as an e-book (excuse the pun because it is not that natural).

What would you like to do to collate your experiences into one project?

To have a small holding in the South of England. Having access to modern technology in some areas of it and the other spaces on the land to be without mains electricity, mains water or gas. To be with other people, where we drink pure water, work with earth shelter, cob building, wood, animals and be as close to nature as possible. Make bread in a circular stone oven that we have built – which also heats the building and the water. With a cool cellar and a breezy platform above in the trees. Yes, this is my wish – but if I taste it not I shall worry little as I have savoured each ingredient individually during my time working my way around the world. I have also now written up all of my travel notes and published novellas, one for each country I have had adventures in. They culminate in an epic full size volume, containing all 12 in the series. Here is one part of it to preview now, an in depth wwoofing related journey I made in England, written in *3rd person literary non-fiction*.

CHAPTER 26

A WWOOF JOURNEY IN ENGLAND

Taken from the series:
***'The Adventures of a Greenman
Part 10 - Raw Travel England'***

Once again Adam volunteered with the *Wwoof* movement, first of all, on farms nearby, in Sussex, and then on some of the many others around the country. It was September, a wonderful time to work in the English countryside. There were hundreds of opportunities on organic farms around England and Adam began to plot a course for the coming months. He had left his seafront flat for now and was on the road again and the idea of not knowing how long he would be gone or what may happen thrilled him. He had done much 'wwoofing' over the years and it always made him feel at peace. Through his work, he began to heal the imbalance that he still felt from time to time from a relatively recent relationship break. As he sowed seeds into the land, he also sewed back together the pieces of his broken heart. For when he loved, he really loved and when he lost, he lost. It would not always be so, but the wanderer was still finding his way.

He did not need to go far in order to reach his first destination, for it was just a bus ride away from where he lived. The wwoofer began by helping to restore an old gypsy wagon. At other times he dug trenches for the electric cables that would light this old travellers dwelling. He was helped by another wwoofer from Germany and Adam found that his trench digging was very precise. The simplicity of the work was almost sacred to the wanderer, as he volunteered on farms, in exchange for his bed, board and infinite possibilities to learn about the land. On others, he would learn more about animals, livestock and the food that is grown to meet our basic human needs. Such delight is known to few men in the grand scheme of things, in the fast and frantic world that we live in, one that Adam found it hard to be in. Yet in time, his work on the land would also bring him back down to earth.

Adam still had difficulty processing information, due to being dyslexic, but still not knowing about it. So when his wwoof host began giving him the jobs that she would like done, the wwoofer told her that he would rather she told him no more than two jobs at once, for he could not remember anymore than that. When he had finished his jobs each day, the eager host would rush around giving him long lists of other things that she would like doing, it made little sense really, for the list just got longer and longer and more confusing. Adam mentioned it to the fellow carpenter who was working on the Gypsy wagon and he sympathised with the volunteer and said:

"That is what she is like, I get it all the time!"

The working wanderer could not tell his host what the problem was, because he did not know exactly, but only that he could not understand more than a few words of simple instruction at the time. Usually this worked out just fine, but in this case, no amount of requests from Adam made the slightest bit of difference.

Information also came in the guise of objects, the trees, plants, tools and equipment his host showed him, they would also have to be understood. They were not complicated instructions, for example one may be:

"Take the saw hanging up in the shed and cut this branch off for me please," his host may ask.

Two things struck Adam about this simple instruction, firstly, he could assimilate information about objects which were not man made, easier than he could things that were. Nature had a natural order to it and the objects in it seemed to be less disruptive to the eye, Adam could assimilate these far easier than he could objects which man had designed, for example, looking at tools in a shed, which would just be a blur to him. Both types of information would still take the wwoofer considerable amounts of time to understand, particular if there were more than two commands, but on the whole, volunteering on farms was much simpler than being in urban areas, where Adam was overwhelmed with information which his brain would not compute. The following year, it would finally make the connection and discover that he is punch drunk, stemming from being repeatedly bullied and knocked unconscious as a child.

From Sussex, the nomadic farmer moved across to Dorset and helped round up sheep and cattle for a few weeks. Soon, his fear of beasts such as bulls and mad cows began to dissipate, as he became familiar with techniques to over come it. He realised that often the animals were as frightened as he and that their main concern was whether or not you were bringing them food or you wanted to make them into food. If a bull was not threatening him as he fed them, he was calm. Once he had learnt that if a bull was threatening him he must still stay calm, but could then jump up and frighten a charging attacker. Bulls knew then not to challenge the unpredictable helper.

It was common for volunteers in the wwoof movement to stay for just a few weeks and then move on to another place, though sometimes they may stay for a far shorter period or far longer, for months or even years. It all depended on what had to be done, with regard to both the wwoofers plans and the hosts. In time, Adam too would find a particular region that he was very fond of and have the joy of settling for a while in one place. Dorset was beautiful and Adam moved on to another farm, where the main help needed was in the woodlands with a small team of people. The farm was just a small part of the enormous estate that was owned by an English gentleman. Adam stayed in a stable that had been converted into a cosy little dwelling and got straight into the tasks at hand, with the other land workers.

The sun cut into the forest as they walked through the delight of December. 'They' were three. One wwoofer and two foresters, heading for an area they would tidy for the day. As they walked, frost under foot, they spoke of trees. Pausing by a Yew tree, one told of his love of making longbows, an astounding ability to make such ancient weapons with skills nearly lost. He would bring in his War Bow the following day and all would marvel at it and the whooshing sound as the arrow was released. Yet for now, they worked, using only hand tools to cut back an overgrown part of the woods. It was soothing to be in nature again and the nomad was in his element. Working quietly, far from the noise of busy roads and modern places. The calm of the activities soaked right through Adam's hungry soul, as he worked away until his heart was content. A fire stood in a clearing which had made and they all watched it silently, one of earth's great attention keepers, the wanderer is thrilled by such natural wonders.

As he is placing leaves that he has collected on a compost pile, he sees that an acorn has begun sprouting in the warmth of the decomposing process. He has never seen such a thing and he marvels at the beautiful small seed of a tree that he holds in his hand.

One day, this may be a grand old oak tree, he thinks to himself

Nature is incredible, I cannot believe that such a thing can happen.

To him it was no different from alchemy, this tiny piece of matter that could have easily just dissolved into the ground and disappeared. Yet when the conditions were right, the energy it stored within had sprouted out and with proper nurturing, the seedling would flourish. He asked the woodsman if he could have the small life that he held and he agreed, happy that he saw the magnificence of the simple thing, as they too loved the wonders of the woods. He took the acorn back with him to his room and planted it in a very small pot, watering it and watching it grow, little by little.

When it is time to move onto the next venue in Dorset, Adam hops on a bus, with his backpack containing the oak seedling and makes his way down to the community where he will stay. Like the first place he ever stayed at when his travels really began nearly ten years before in Brazil, this one also offered courses of a spiritual nature. It is because nature is spiritual that they also give workshops on the more practical matters of life and the land. They have one on making a sewage water filtration system from reeds. Another on repairing garden tools and another on Hazel Wood Coppicing.

It is in the grounds of the big old house that Adam helps, tending the garden and preparing it for the depths of the winter to come, clearing the land and enjoying the clear bright days that light up the ice that is all about them. There are many people living in the community and the wandering land worker enjoys there company, until he begins to fall ill. Adam develops full-blown influenza. He receives the soothing company of a lady whom he has met in the community. She visits him, checking his fire and making the shivering patient hot drinks. She seems immune to his flu, until a few days later she too begins to ache, but the strong girl just carries on tending the land and soon her symptoms go away. Adam tries, but it had hit him differently, this was not a case of 'man flu', but we all have different constitutions. It is immensely painful to even stand up, he feels completely hammered all over and so he resigns himself to just receiving the unconditional help that this lovely soul administers to him. This is the force of nature, a virus so powerful that it can overtake one's own desire to move. Not even the nomad can go anywhere now. He is lucky to have such an earthy girl to help him and he thanks her by giving her a gift. He told her the story of how he discovered the acorn and what a wonderful thing it seemed to him. Adam then gave her the small seedling, which he had carefully carried from his last host's estate.

One week later he is up and about and back to his duties in the community. Once again, the nomad lives with the bliss and simplicity of open fires, chopping wood and being amongst like-minded souls. Though two of the community members are a little bit too like-minded as they also like the lady that Adam is beginning to form a relationship with. Only, she does not want to be with them and is fond of the wwoofer from Sussex. They snatch the odd moment they have off from work to roam in the countryside and be away from the jealous men. It will not be long before the wanderer makes his way to his Christmas placement, a woodland farm in Devon where he has organized to stay.

The nomad bids his lady friend goodbye for now and takes a train to Devon. He buys a return ticket as it is only a few shillings more than a one way.

Strange though, he thinks, he has no plans to come back this day. Yet on the way, he does not get such a good feeling about his next farm host in Devon, but not sure why, he continues, hiking the last five miles through country roads to get there. As he arrives, the place looks deserted, there should be at least a few people about, according to the details that he has received, but there is no one there to greet him. Later, he will discover that they had forgotten that Adam was coming. He is not sure what to do now, Christmas is not a time to be out alone. Perhaps he could return to Sussex and stay at his old flat, but instead he decides to ring the community that he has just come from. After explaining what has happened, they welcome him back to stay with them for Christmas and the lonesome wanderer is relieved to have company at this time. He makes his way back and settles in.

It is in a morning meeting the following day that the gentle soul receives an unpleasant shock. One of the men, who has a liking for the lady Adam has met, becomes outrageously rude. He is so annoyed that his territory is threatened that he throws insult after insult at the keen volunteer and because this man is a community member, who has expressed a desire for Adam not to stay this Christmas, the nomad has forty eight hours to leave. The group had agreed that they would not receive volunteers or even guests in their Bed and Breakfast over the Christmas period and though all but he and the other lady fancier are the only two who object to Adam's presence, the group must stay with the decision that they have agreed upon. It is a shocking state of affairs and one that several others of the members are appalled at too, as they are only too pleased to have the company of the friendly wwoofer over the Christmas season. He finds himself somewhat distraught at the man's vicious attack. He consoles himself with the best friend he has, nature, doing what has to be done. Wandering through the country lanes, he makes his plea to the Oneness.

Dear Oneness,

I find myself in a bit of a predicament, would you please guide me to where I should be?

Few would believe the events that follow, they are certainly going to appear in his nomad's notes soon. A particular farm came to mind that Adam had read much about, a small holding that had a legendary reputation that was not far away. Several times, Adam had tried to arrange a time to stay and help there, but it just had not come about. He knew that it was but a stones throw from where he walked right now and decided to take a chance by paying them a visit, they sounded like a friendly bunch of young organic farmers. It was worth a go, perhaps they would be in need of some help.

Ice underfoot and weight in his heart, he followed the country paths of Dorset until he came to the place – not knowing what to expect. As he arrived he was greeted by the scene of a team of people preparing a cider press and the warm call of "hellos" and "come in". He explained what had happened and that he was looking for a place to wwoof over Christmas. It transpired that the farmers had hoped to go away, but had not managed to arrange a 'house and farm sitter' so one of them may have to stay behind to tend to the many animals.

"Yes, I remember speaking to you on the phone Adam, thank you for coming over to visit us. Look, it might be a bit of a lonely time for you here on your own, but you are welcome to stay and look after the animals if you like. There are loads of pigs that need feeding! Then I can go away with my family for a few days. What do you think?"

Having had quite a bit of wwoofing experience in dozens of places, he jumped at the chance to look after what he could see was a beautifully kept little farm. He could move in to a caravan that they had immediately, and although temperatures were around freezing, it had a neat wood burning stove to keep him warm. Adam would have a few days training on the essential needs of the Jersey cows, pigs, chickens, geese, ducks, dogs and cats, then the family would go off for Christmas to see their parents. It was a wonderful opportunity, the farm was absolutely idyllic.

In the end, it would not be such a lonely time, for Adam's female companion would come and visit him often. There would also be other visitors who came to lend a hand with milking the few cows that they had. They too stayed, sharing the rich and tasty pheasant stew that they had brought with them. Yet each morning, the task of feeding the pigs was down to Adam. They were beautiful animals, not just common pinks but spotted and shaded with browns and blacks along their backs. They seemed to be incredibly intelligent animals, the nomad would need to be on his toes. Some were big, the males especially could be particularly intimidating, if one allowed them to be so.

It was a thrilling time for Adam, he could not want for more, so close to the type of environment that he had been born into himself. Yet with every beginning comes an end and this situation was no different. The wandering pig sitter heard news from the community up the road, that the insulting man who stayed there had been asked to leave, he was no longer welcome at the place. Justice had found its way of rising and Adam could not be having a better time these days. He was still friends with the lovely lady he met and from time to time they would go out on an excursion into the big wide world. Romance would not blossom, but one cannot make things grow in life that are not meant.

When the owners of the farm from heaven return, they invite Adam to stay on longer. He goes to the local market with them, selling the produce from their farm and attends social gatherings that they host in their rather splendid barn. It is staged on two floors and the sheer volume of guests and cider soon raise the temperature of the place. This is truly a phenomenal Wwoof venue, the depiction of the ideal opportunities available on organic farms around the world. Adam has come to love this way of life and all that it offers. This particular farm is hugely popular and the nomad reminds himself that when he writes down his nomad notes, that he will not name the farm, for they already receive hundreds of requests from people who would like to volunteer there. He has hopes of publishing his thoughts on helping on farms and would not like to burden this venue with many more enquiries, for they would never get any work done in this case.

Adam had already started to make a few friends in the area, one had a farm of his own just a half a mile away. He too has many pigs who roam freely and invites Adam to come and help on his spread of land. There is a simple place to stay on the land, it looks like a gypsy wagon, but is actually a shepherd's hut. It would turn out to be the farm volunteer's most memorable wwoof accommodation, a tiny wooden home perched upon an old trailer. Steps led up to its creaking door, with ample gaps in it to ventilate the wood smoke. It had been made by the farmer, not having a house on his land, he had lived in it for a time, before upgrading to a monster of an antique caravan, that was equally attractive. A bent old chimney took the smoke away from this romantic but practical dwelling, set amidst fields – where pigs wandered nearby. Water was from a hose pipe outside, unless it was frozen up, for the depths of winter had set in.

The toilet was of the kind that Adam had seen many times before as he travelled far a field. All of the farms in France, Portugal, Spain and Hungary which he had stayed on used this compost type depository. A tall timber room with a section below to collect the essential fall, the throne often having unrivalled views of nature. Often mixed with ash, soil or straw, but in this case there was ample saw dust to cover the daily needs. Utterly hygienic, a system that allowed its waste to decompose in a short period of time. It was a peaceful place to be, looking out over the field of hungry pigs from a frosty seat.

Having been in the area for some time, Adam got to know a few more local people. One asked him if he would restore an old wooden bench that she had and the carpenter at heart was only too pleased to complete the work once his daily volunteering was completed, happy to add a few gold coins to his purse, for it was running low. When the job was done, the lady then asked him to make some things for her out of wood and before long, Adam had enough money to consider fulfilling his plan. For he had wanted to Wwoof from Sussex to Scotland, but such dreams still need funding, even if his lifestyle had little financial outlay involved. He wanted to experience as many different farms as he could and acquire a varied range of skills. Though for the time being, there were other adventures to be had where he was, until a natural gap appeared again and of course it always did.

The freezing water pipes presented a few problems with regard to getting the pigs the water they needed and when it came, they would make the most of it. Adam had left it running to fill their troughs whilst he fed the chickens that roosted nearby. When he finished he turned around and saw that the trough were full, but one of the brighter pigs was dragging the hose in his mouth, over toward an large dip in the ground, that formed a pond during wetter times. When the smart swine reached the area, he dropped the hose in to let it fill. The nomad did not have the heart to turn the water off until it had filled, by which time, several of the pig's friends had arrived to splash about in the icy water. Adam's father had often told him just how incredible these animals were, as he had been a pig farmer, now he was seeing it for himself first hand.

A little gas cooker gave a hint of luxury in the hut where Adam stayed. Normally, he would eat in one end of the barn that was the other side of the farm. A room had been built in it, with a powerful wood burning stove, this is where visitors gathered and food was to be found. At lunch time Adam would eat simple meals of pork and bread, sometimes cheese. The evenings would see a wide array of winter vegetables that were grown on the farm, often in polytunnels, but the most attractive part of the nomad's diet was the endless barrels of home made cider that stood under the eaves at one end of the barn. When the wwoofer tired of such simple life, he wandered to his friends that he had made at the Christmas farm, where he helped them make sausages and shared a bite to eat with them.

Life was so different from that which Adam had known in the summer months in Sussex, where he had been diligently writing about his travels, such modern things as internet and television he rarely saw on his farming adventures. He was so tired from the days physical work here, in the elements and air, that sleep was long and deep, broken only by the call of the empty wood burning stove in the shepherd's hut, at which he would rise in the night to feed it. There was always something to feed on the farms, even the small bird that would come through a gap in the roof of the shepherd's hut, he too was after a morsel of food. His visits became more frequent when snow began to fall. Most people would not like a bird flying into visit them in their house, but Adam's situation was rather becoming to the visitor.

In time, the nomad was becoming more and more a part of the land. His hair had got wild and curly again, as too had his beard, he was at one in the natural environment, until relentless snow storms battled. Initially, there was no problem with the snow, the wwoofer was warm, fed and occupied, but when the drifts stopped all of the other activities on the farm, the place became a difficult one to manage in. Everyone in England was suffering, it was January 2009, the snow was so thick that the country was coming to a stand-still. The army were called in to many parts, to clear the roads, people were getting stuck all over the place, stranded in the freezing conditions. Yet some of the main roads were still passable, one of which was only two miles from where Adam sat. He made a decision to return to the flat if there was still a space available for him there. A phone was the limit of the farms technology and Adam made full use of it by contacting his friend who owned the seafront abode, the wanderer had done without a mobile phone up until now. To Adam's delight the flat was free and he wasted little time in getting too it. The ground was rock solid where he was and not much work could be done amidst the blizzards that covered it. He looked forward to the comfort of brick walls again and the luxury of being able to turn on heat by the flick of a switch. To have tap water twenty four hours a day. The central heating lacked the wonderful whiff of smoldering logs, but the wanderer was ready to forego that pleasure and replace it with the ease of which his abode could be heated.

A few days back in civilization would be good, he would have a chance to arrange some more wwoofing placements on other farms, hopefully working his way all the way up to Scotland. The snow was still thick, but Adam felt that it was a good day to hitch hike, as he was far more likely to get a ride on a day like this than any other. People did not often stop for hitch hikers in England any more, they were too frightened that they would get attacked, and sometimes they were. Whoever picked him up today would be brave enough to be out in the fraught conditions and tough enough to deal with anyone thumbing a lift who got out of hand. Adam stuck out his hand and looked like someone who really needed a ride. A sensible looking chap, even with his slightly wild look, but to the trained eye he was clearly a man of the land.

To his delight, he did not have to wait long on this snowy day, it was a young man in a van who offered the ride, as he pulled into the lay by where Adam stood. They got on well and talked about travel and the wild. Adam noticed that the driver was wearing a *National Trust* sweat-shirt, this was a warden who worked for the countryside management organisation. Adam was right, this was a man of the land and he had spotted the nomad as being one too. He would have to get four more rides and walk an awful long way before getting back to Sussex, it would turn out to be a terribly long day in the snow, however, he would make it safely. It would be a good day to hitch hike and the journey would be free, all except for the lunch he would buy his current driver, for the nomad was so pleased with the news that he was about to find out.

"How do you get to wear one of those then?" Adam asked, remarking on the drivers proudly worn emblemed sweat-shirt. Perhaps it would be his destiny to become a warden?

He found out that he could volunteer for the group and that they were actually looking for an assistant warden in Devon quite soon. This was an incredibly rare opportunity, many people longed to be a warden for the National Trust. It would not be a post that he would stay at long, for he was not cut out to be a warden. Yet it would be a phenomenal experience, one that he had to have in order to know that it was not for him.

"Do you mean I will get to drive one of those lovely green Land Rovers too?" Adam asked, sure enough, he would. A week later, he would attend an interview and secure a position. It would not start until May, but that would be fine, he would have time to carry on wwoofing for a while, before making his way back down to Devon to volunteer in a green Land Rover. It would be another experience and another string to his bow.

Adam arranged another couple of farms to volunteer at whilst he was back, that would be enough to get him on his way. There were infinite possibilities to learn about matters of the land and all that lived and grew on it naturally. There was no need to pay out vast sums of money for courses if one had enough time to go and wwoof. Adam learnt that the lady who started the movement, did so in the same year that he was born. Initially it had been called 'Weekend Workers on Organic Farms' and was aimed to help people who just wanted to get out into the big green to help for a couple of days. It grew so widely, that in time, helpers could go to nearly one hundred countries around the world, experiencing a cultural exchange, whilst volunteering just four to six hours a day. The group interested Adam hugely and he began writing down all of the practical things that one would need to know in order to go and wwoof. Though there did not seem to be one central organisation that ran the wwoofing in each country. He alone had wwoofed in five already, since he began his journeys nearly ten years before and he was about to add another two to the list as he worked his way up England, onto Wales and Scotland. When he did some internet searching, he found that there was not a book on the subject and thought perhaps he would write one. He discovered that each country ran their wwoof group autonomously and realised that it would be impractical to contact each one to ask if they objected in any way to him writing the book. He also decided that he would not look at any of the guidelines to wwoofing that appeared on websites, but instead devise a list that he found had been useful to him. Ultimately, it would turn out to be very similar, yet it was spiced with humour and tales of his own experiences whilst he had wwooofed and worked his way around the world. Adam did not know it yet, but he would meet the founder of Wwoof and present her with his dinky little pocket book. She would be very pleased with it and it would go on to sell in many countries around the world. It would also lead Adam to new frontiers, as he would have to embrace the tools of the modern age, such as computers in order to produce it. Yet the few shekels that he earned from the sales would help him to have other land based adventures.

Adam is drawn to help on a farm that works with Horse Whispering. It has always been a dream of his to buy a horse in foreign lands and ride through wild places. It may not be a dream that he chooses to make a reality in this lifetime, though he is still keen to be close to the gentle art of Whispering and makes his way up to the farm in Wiltshire.

When he meets the man who will show him this way of communicating, it is no surprise that he has a peaceful demeanour and is quiet, confident and serene. They walk together, slowly to the horses and the man stops some distance from them. A stallion comes over and nuzzles him. Adam is a little apprehensive around animals he does not know, but wisely so. He is soon at ease in the presence of this beautiful horse, watching the Whisperer, looking at his energy, which he sees visibly radiating. It is clear that the horse also feels his peace. Movements in each of the communicator's body language are subtle, but the slightest change can reveal much to the eye that sees intention – and Adam's is becoming finely tuned.

He is invited to stand next to the Whisperer and Adam learns by 'direct transmission', a technique he remembers first seeing from a martial arts teacher in Spain. The teacher consciously transmits physical energy in a non-physical way. Thought to thought, mind to mind, only in this situation, the whisperer is doing this with the horse, not only sending his own intention, but also listening and reading the horses. It is a powerful tool for the receptive, who may sense the intention of the next action or movement. This is very useful in martial arts, but also to the horse whisperer. Animals are already well developed in what is not an art for them, but more of a sixth sense. A student witnessing such events may absorb knowledge rapidly and become extremely sensitive to all that is happening before his eyes.

The highly interested wandering wwoofer is permitted to go inside the enclosure with the horse. They approach each other slowly, smelling, looking and watching. When the horse tries to win the partnership and be ruler over Adam, he submits at first to the animal's grandeur and looses his footing. The Whisperer encourages him to make back his ground and be in control of the horse, letting him know that the human is chief. It is an unnatural state of affairs, which only came about when man took ownership of such beauties, but as that is the reality here, Adam expresses his leadership. In no time at all, he has the respect of the beast.

He may never learn all of the simple rules and ways of Whispering, for it is not his path, yet what he has seen will have a knock-on effect with his understanding of man's relationship with other living things. Whether a boy who bullies viciously or a person who speaks to him maliciously – the rules are very similar. Energy will affect us if we give it the chance to do so, it will get into us. In the case of the horse, it will not last long. For an owner, it can last a lifetime, if one develops an uncomfortability around the animal. It is of course a natural law. The threatened must rise and show that they will not be beaten, rise to a place of seniority. Though in time we may find that we do not need to fold to the feeling of fear, of being beaten or the insecurity of not knowing another person. It has taken Adam a long time to over come his own fear of being hurt, after his head has struck the ground several times in bicycle accidents and fists have also knocked him out.

Man has brought animals out of the wild and tamed them for his own use. In doing so, we no longer have many beasts roaming our lands and have ruled out potential threats around us as humans have evolved. Though naturally, large animals still have their primitive senses and finely tuned perceptions, as perhaps humans do if we dig deeply enough within ourselves. It would be a while before Adam would be tested by life to the full, to see how much he has learnt about these things and in time he would face a deadly challenge. His life will depend on drawing upon his knowledge.

For now, he will face another test. On leaving the horse whispering farm, he moves onto to his next placement on the Welsh English Borders. As is the law of averages, not all we meet will be compatible with our own psyches. We can keep quiet and bare it, or if possible, remove ourselves from that which does not fit with us. In this case, Adam chose to do the latter, as he is not suited to the unpleasant temperament of the owner of the next farm and is unhappy there. Subtly he tells the owner that he will not be staying. Yet his next place is a week off, he must find somewhere to sleep in the meantime. Though he has his train tickets, Adam has less than £50 and no savings, he has no credit or cash point card anymore. The days are bright, but the nights are icy and he relishes the thought of a challenge, making a shelter in a thick woodland for the night. The going is hard and he is tried to his limits, as he feels like a homeless wanderer, but a fire soon warms his spirit and spurs him on through the night.

When daylight comes, it is with bright sunshine again and though it is still freezing, the wanderer is happy to be free, glad that he did not stay on at his previous place. Toward lunch time, he reaches a pub and goes in for a pint of their finest ale and a jacket potato. He has a plan, for he now travels with an old mobile phone that he has been given and what is more, he has credit on it. The beauty of the wwoof world offers the solution, for he calls some farms to see if they are in need of any immediate help for a few days. The first two phone calls brings no reply, the third an answer machine, the fourth a "not at the moment thanks", but the fifth brings a 'yes' and he is off again, hitch hiking up to North Wales. It is a tiring business travelling in this way and the nomad is relieved to reach the security of a farm again. A small place, but the hearts of the owners were big. They welcomed him warmly and showed him to the caravan where he would stay. There was a heater in there, but thankfully the weather had made a turn for the better and things were starting to warm up a little.

In return for the hospitality they offered, Adam did what ever tasks needed doing. They mostly grew vegetables in the two polytunnels, so the wwoofer helped to weed them through. Though by the grace of wwoofers-luck, they also kept a few horses and the helper was able to get involved with feeding them and cleaning up all the leather tack that they wore. The host's two children were also happy to have some company and Adam enjoyed playing football with them in the ample stretch of grassland that they had made their pitch. It was not a bad way of life to have, for the nomad or the children and he remembered his days on the small farm that his parents had. In some ways, he was reliving his youth, but in others, he was doing something that he could not have done then, at such a young age. Now he was learning new skills, for the road ahead would take him far a field again and it was important that he was comfortable with being out in the wild. Though for now, he prepared to move on from this very short stay and continue his wwoofing journey in England, up toward the Lake District of Cumbria.

The variety of places that Adam had stayed at over the years was enormous, this time his bed would be in a bunk house, accommodation used for groups that visited the Lake District and stayed on the farm. It was a wooden building, probably constructed in the 1920s and was big enough to house a dozen or so people, but he was the only visitor at this time. A couple with their two children ran the farm, a six hundred acre place that stretched way up into the crags that the farms sheep roamed upon. It would be Adam's job to go up into the mountains with sheep dogs to bring down the vast flocks of sheep for their immunisations.

On the second day, after the wandering wwoofer had been shown around the lie of the land on the back of a quad bike, he went up to the fells with a few farmers to round up the sheep. Adam's job was to cut them off before the scitty animals made for their escape. It was not the first time the wwoofer had done this, on a farm in Dorset he had had to round up Hebridean sheep, whilst they tried to jump over the top of him. This variety were slightly more subdued, but it was the boggy wetlands that he had to cross that proved to be the most difficult work, running miles from rock to rock, as he tried to head off the sheep and send them back down to the farmers. It was exhausting work and took all day, but the nomad enjoyed the challenge immensely.

All the sheep were brought down safely and the next day, the farm help learned how to carefully turn them onto their backs whilst they had jabs for all sorts of pesky mites. Although it was not the right time of year to shear the sheep, Adam also learnt that it cost more to do the job than the farmer was getting back for the sale of the wool, in some cases it was barely worth selling.

How did farming come to this terrible state of affairs? He wondered.

There were also horses at the place and the nomad's fondness of the animals got him the job of feeding them, along with bringing in the chickens at night and digging over the vegetable patch. These were years that the wanderer could simply never forget, the tiring physical days and the early nights, often with a pencil scribbling away, adding to his nomad notes.

When Adam went out way into the depths of the fells with the farmer and his truck, he learnt that he was only a few miles away from his next host, who was also in the magnificent Lake District. The next time they went, the nomad took his back pack too and dropped it off at the farm. He was not due at the place for a few days, but decided that he would walk across the mountains to get there, as it was only about twelve miles away and would be much easier without his pack. Sometimes the wanderer was exceptionally well developed in matters of the land, other times he just did not seem to grasp basic principles at all. For although his next wwoof host was only that distance away, twelve mountain miles can be like a twenty five mile level road walk. Nevertheless, when it was time to head on over to his penultimate farm on this tour, he went on foot. Once over the peak that separated him from Lake Coniston and the next farm, he popped on a bus for the last leg of the journey, arriving at the breathtaking watermill where he would volunteer for the next week. Although the wheel was not running, it was still a staggering sight to behold, because of its height and sheer size.

It must have been the quintessential organic farm. A river running through it, the hosts with volumes upon volumes of knowledge stored in their minds and hearts and an exceptionally well organised schedule for its wwoofers. There were two this time, Adam and Caroline. She was from London and would partner him in his tasks through the week. Henry, the owner, turned out to be a director of the Wwoof U.K organisation. It was a complete surprise to the wanderer when he spoke about his plans to write a simple A-Z book for wwoofers. Henry then told him about his own connection, which he held in a voluntary capacity. He loved the idea of a practical book about wwoofing and when Adam had finished writing it the following autumn, he sent one along to the warm host.

The mill itself was full of antiques and bicycles, which Adam found to be incredibly interesting. There were cycles from every decade and perhaps an antique for each one over the last five hundred years too. It was like walking into a fairytale for the nomad, he loved all of these things, particularly as he was often able to admire them whilst he broke off from making kindling wood next door. As with many of the farms, the home cooked food was delicious and these particular meals were cooked in the Rayburn oven that Adam so often split kindling for.

Working with wood was a favourite activity for this host, during the week they made a combination between a gate and hurdle from green timber and Adam thought perhaps that it should be called a 'Gurdle.' He was becoming increasingly interested in words, as he spent more and more of his free time writing. It had been a joy to tour some of the farms on England's green and pleasant land. People would often ask him in the future, what his favourite country was that he had been to and his answer would be *England*. Though soon, he would head off to Scotland, for his final farm stay this year. The adventurous walker decided not to try and walk to it, although it was wholly achievable, he had already secured a discount rail ticket to Dumfries. From there, he would hitch hike to the place where he would stay.

He reaches the Scottish farm safely, after exchanging stories with yet another driver who picked him up. This man tells the hitch hiker all about the woman who he is separated from, who now lives in the middle of France and how he drives over to see her in the truck that Adam is in now. It is a 1970s Land Rover, the ride is as rough as the surface of the moon and the hiker is glad that he is not travelling that far with the gentleman. It is a strange phenomenon, getting into vehicle with somebody you do not know, likewise, a vehicle stopping for somebody they have not met before. You are suddenly a part of each others lives and the unsaid deal of chatting openly is nearly always done before the driver even pulls over to pick you up. Then as quickly as it began, the relationship is over.

His new hosts leave most days to go and run a shop in the next town. It is a grey and lonely farm which Adam finds himself on, but when he has finished his tasks, he spends time drafting the book that he would like to publish. The nomad has now volunteered on organic farms in England, Scotland, Wales, France, Portugal, Spain and Hungary. Doing other voluntary work in Brazil, Latvia and India too. His concise words describe how one may go about such a thing, written in a brief and clear fashion. The same year, he will publish it and it will steadily sell a small number of copies worldwide. He can see this happening and the goal keeps him feeling bright whilst he works mostly alone on this damp and windy placement in Scotland.

The wandering wwoofer does make a few friends at the farm, they are baby goats whose mother has died. His job is to hand rear the kids by feeding them with milk from another goat. Each morning he takes out the five bottles of milk, which has been warmed again to a mother's body temperature and tries to get the sweet little monsters to suckle. They always take the feed, but not always straight away. It is a wonderful job to be given and Adam willingly attends to the babies each day.

Sometimes the job is followed by one of feeding some bulls too. They are young, but still well able to cause injury if one gets on the wrong side of them. Adam makes sure he stays on the right side of them. They are a good natured bunch and soon get used to the idea that he is a bringer of food and so they do not bother him. The wanderers confidence in dealing with animals has risen dramatically since he began wwoofing, in turn, the practice has helped him become as focused on earthly matters as he has previously on that of the esoteric. For a time he will sway nearer to being in the body than he will to that of looking into the unseen energies of life, that have previously been of such great interest to him. It is a leaning that he will ignore for a while, until he learns how to balance the two.

When the farm trainee is not feeding animals, he is digging trenches again, but this time not for electric cables, as he did in Sussex, but for potatoes. The farmer's wife has some days off and shows Adam a simple technique for increasing yield for the crop. As it grows, each time the plant appears above ground level, one can recover it with fresh soil. This way, the plant stem sprouts out again and these arms will also eventually grow potatoes. The procedure can be repeated a few more times until by cropping season, the trench is full of the goodies. These types of facts thrill Adam, he is so interested in the simple things in life, the ones that sustain the basic human needs. How is it that a child can know how to reprogram a computer or download things from the internet, but not know how to grow a potato or make fire?

Because they do not have too? Is that the answer? That is why matches and shops were invented is it not? Perhaps, but it is tragic when as they grow up, they do not even know how to light a match, or find their way to a shop without using a GPS system. Was man perhaps forgetting how to be one?

It will soon be time to go to Devon so that Adam can take up his voluntary position with the National trust as a Countryside Warden. He makes his way from Scotland, back down to Sussex, where he has secured a few weeks building work. Staying in his flat again, overwhelmed with the comfort he finds there. The place smells unlived in, it is not often used and in the short time that he is there, he gives it a thorough airing. The wwoofer is very lucky to have the use of such a place, with a very favourable arrangement. Adam had house-sat all over the world, it just showed that similar arrangements could also be occasionally made in England too. He also has the joy of seeing some of his old friends again, although in the years that he has been away, he has lost contact with many. It is spring 2009, the road has been long since Adam left for his wild adventures at the turn of the millennium. His brothers and sister, mother and father all live nearby and when he calls on them, they are pleased to see that he is happy, though it is not long before they bid him farewell again as he leaves for Devon.

The money he has earned doing building work, repairing the outside of a house, will be very useful in the coming weeks, for the new position with the National Trust is unpaid. As with wwoofing, Adam will receive accommodation, however, he will not receive any food or money toward it, he must find his own way of providing this. He worries not about the arrangements and instead focuses his attention on getting down to Devon. Buying an old bicycle, he fills it back and front with things he will need for his stay in the countryside. Hopping on a train with it, he makes his way on yet another adventure, open to all that it may bring. Strapped to its side is an air rifle, he is prepared for the possibility of leaner times ahead.

The cottage he stays in is wonderful, a gorgeous old place in a quiet sleepy village.

"It should be a bit more lively when the other volunteers turn up," his new boss tells him, but what Adam does not know is that they will never come whilst he stays there, for his boss has not arranged any yet. Nevertheless, there will be ample to interest the new land worker, there are paths to strim, paths to strim and paths to strim! The absence of other volunteers means that Adam must do most of the work. Whilst he does it around the wild places backing onto the village, he notices that many tourists visit this quaint little place. It seems that there is nowhere to empty their purse into, or buy mementos of their country day out. Adam's funds are becoming low, soon he will not be able to eat and he takes a chance by investing his last shekels in something that he thinks will make him some money. Flour, he will make rustic bread to sell to the visitors.

There is water in his tap and Adam knows how precious it is to have such wonders. He had once learnt how to mix it with flour to cultivate a natural yeast, known as Sourdough, that is ready within a few days. By regularly adding to it, the yeast may continue to be good for years and years. A warm spring is upon them and Adam decides that he will turn his hand to making Rye bread and bakes many batches of it. Early each morning before work, he bakes the loaves that he has prepared the night before, places them in a basket outside the cottage and sits it on a table, covering them with a cloth. He places an 'honesty money box' beside his sign that reads: *Rye Bread for Sale*

Each night he brings in the basket, that is usually empty and with it the money box. In this way, he has just enough cash to live on.

The weather is hotting up and so he makes a simple lemonade and sells that too. He is pleased that his small venture works and through it makes many friends amongst the villagers, who also buy his goods, as well as the tourists. In the garden that comes with the cottage, he plants potatoes and other fruit and vegetables, using the techniques that he has learnt over the years. Around the paths that are his job to clear, wild garlic thrives, so Adam cleans it and puts it in jars with olive oil, this too he sells outside the cottage. Strictly speaking, he is not permitted to sell anything there, but all turn a blind eye to the wanderer who is making his way in such a resourceful fashion, even his boss kindly looks the other way. He could be claiming Unemployment Benefit, as volunteers are allowed to, but he is not. Somehow, the nomad always gets by.

In the summer, he will go to the beach, which is not too far away and catch some of the many Mackerel fish that swim just off the shore. Cooking them on the stones as he did with the fish he caught in India and remembering much about his travels as he does so. In his spare time, he works on the book he is writing about the wwoofing and volunteering he has done on his way. It is all done on paper, as the wanderer is a little behind the times with regard to technology and he does not know how to prepare it on a computer. When the draft is finished, he sends it off to his sister in law, who will kindly type it all up for him so that he can go to the next stage of having it made into a book. Who knows, perhaps it will make a few pounds to help him on his way, for sometimes he does not have any money at all.

Things become tough when rainy weather sets in, as he cannot put out his goods for sale, but he must still eat. He is living on a shoestring and if he is not careful, he will have to eat his own laces. The rain goes on and his money runs out completely and for a couple of days he survives on wild garlic rye bread and lemonade, but man cannot live on bread alone, he must find something else to eat. As he looks out from the bathroom, onto the garden, he sees a rabbit hopping about. Adam picks up his gun, leans out of the window and shoots his dinner. He skins it and keeps the fur to make a lucky purse with, once it is dried and cured. Adding some spices to the meat and two cups of rice that he has left, that evening he enjoys a tasty meal. Though he does not like to kill, he had to eat and the nature of man rises up within all who hunger. He invites a friend for dinner, who brings a few red peppers that he has grown and they too go into the pot. When his visitor pulls out a bottle of wine too, Adam is somewhat overwhelmed, when one has hungered and not fed, such delights are more appreciated and to the nomad, it is as if they are living like kings.

The lucky purse brings with it fine weather again and the rustic baker makes a basket of bread to put out. Soon he hears the sound of coins in his pocket, yet despite receiving not a penny from the National Trust to eat with, some people he works with complain to Adam about his selling of drinks and bread outside the home he lives in. It is not a house open to the public, but because it is owned by them, it is forbidden to sell a few loaves of bread on a table outside. They also complain about fish being barbecued on the stones of the beach, as the law forbids it, but even some of the local fisherman cook theirs there.

He can bear all of this and even the perfectly straight paths that he is asked to make through the countryside, though there is little natural about them. Yet when his boss asks him to make some small flags and go and put them in all the dogs mess on the paths to alert the dog's owners to the peril, Adam at first thinks he is joking, but no, he is serious. It is too much and after trying out the voluntary post for some months, the wanderer realises that he is not cut out to be a warden. Nevertheless, he has driven his green Land Rover and been given a good pair of leather boots, these alone will take him far, to many more places.

It is better to try and know that you do not want a thing, than to know that you have never tried but wish you had. Some of the locals are sorry to see his smiling face and light hearted ways vanish from their village, but he is gone...like a butterfly in the wind onto other shores.

Remarkably, the free spirit has a clean slate again, nothing binds him and he begins to consider new adventures. Will he ever tire of wandering? Perhaps. Though for now, he has a little time on his hands again and asks his friend who has the poorly house, if she would like him to do some more work on the property. It is a massive undertaking and the nomad come builder is glad that he did not try to tackle it all in one ago, alone. The master builder that he once worked with, erects scaffolding around the house so that he can finish another large wall. It is a lot of work as the home needs completely refacing. The owner has afforded another stage of the work to be completed, but the last section will have to wait until another time.

Adam spends weeks on the house and is relieved to have completed this part of the job and painted it all. It is time for a change for the wanderer, for he will wander no longer for a while. He has things to learn and jobs to do. Primarily to find out how to use computers. He decides to base himself in his home town for the summer of 2009, in order to carry out the task and rents a small room in a student house. They are away on leave and the arrangement suits him well, for the first time in years, he is independent and making his way. Adam feels good about this and his raised self esteem helps him to focus on the task at hand, for he has much to learn about computers. He cannot attend a course in such matters, for he still suffers with a serious learning block, information simply just does not go in when taught in the traditional way, ever since he was smashed in the head, and perhaps even before. Sometimes a page on paper or on screen is just a complete blur to him and appears as a sold block of text. He is embarrassed by the condition and is only really coming to accept it now, at the age of thirty seven. Yet he has never told a medical practitioner about the difficulty he experiences, instead, Adam wishes to overcome as much of the problems associated with it as he can himself. He learns by asking the assistants at his local library, each time he gets stuck on some aspect of the infinitely complicated things. They do not mind and in time he actually becomes friends with one or two of the librarians, who like to see his smiling face and naturally happy disposition. He is diligent in the course that he has set himself and every morning he is at the library just before it opens. The maximum time one can spend at the terminals is two hours, then one must leave the machines for an hour or so before one is permitted to return to them. This is ideal for the once wanderer, for he does his shopping or stretches his leg on the beach nearby.

As he learns, he goes through the document that his sister-in-law has typed up for him, the first draft of his wwoof book. It will take several hundred hours more to bring it up to the level it needs to be at, in order for it to be professional enough to sell as a paperback. Yet she has done him a great service in putting his words to paper and he agrees to pay her one day for the effort she has made. At present, he can only type about twenty or so words per minute, but that is set to increase with all of the computer work he will do in the coming years.

Once the computer student has mastered the basics, he learns how to transfer the information he has prepared to an online book publisher. At first, the process seems to be quite complex, but in time, he is as comfortable with it as he is walking twenty miles in the countryside. It takes Adam dozens and dozens of one to two hour sessions, using the library computers to reach this stage, it is a real struggle for the wanderer, the letters and numbers on the screen and keyboard are incredibly confusing and often he has to simply cut his session short because he cannot see straight or absorb the information in front of him. Somehow he battles through and eventually receives a copy by post of his little book on farm volunteering. It will take him five more attempts and the purchase of five more draft copies of the pocket book, before he is satisfied with the content enough to sell it the mainstream market. Before long, he makes it available for retail and it starts to sell at some of the most popular online stores. Not many copies, but enough to encourage the self-motivated student who has come far off the track of the wild to take part in some of the more modern ways of life. He registers with the tax office again, this time his profession is 'author'.

The book carries a disclaimer that it may contain mistakes, as it is of an organic nature. In years to come, these copies will become quite rare as Adam will discover such delights as 'Spell Checking' tools in the future. It is autumn and his pennies are low. The lady he sometimes works for can afford the next stage of the project to be done and he continues the work of restoring the outside of the house that he began the previous spring. The salt air of the coast it sits by has corroded the walls over the decades and he learns the skill of skimming, to bring a little more of the lovely home back to its former glory. He will not complete the project by the time harsher weather comes, as there is a lot of old textured coating to remove by hand first, but gradually the building is taking shape and it is an opportunity to earn a few more shekels to spur the nomad on to his next adventure.

"Nice job Adam, thank you. You are welcome to come back in the spring and do some more work on my house if you like." his boss said.

The more he has worked the property, the more he likes the idea of home, perhaps one day he will have one or even a small piece of land of his own. He now knows enough about how to sustain himself on earth and would like to put his skills to good use. Yet winter is upon him and he does not relish the fact of being in the cold again this year and seeks milder climates. Not entirely sure of what direction to take, he considers what a place would have in order to attract him. A land where the sea would never be far away would suit him well, a sea warmed by the sun. So the wanderer decides that it should be an island that he visits for the winter months, one that knows the winter heat of the sun. The Mediterranean sounds like a place where these things could be possible and so Adam heads for the clear blue waters of the Greek Island of Crete. Not all the people he knows give him their blessing on his journey, for many wish that they could go themselves and wonder how it is that the nomad can live such a life. One thing is obvious, he has no fixed abode and has left the summer room that he had based himself in, stored the few boxes of possessions that he has with a friend and hit the road again. He may be free, but he is also homeless, but Adam knows this and also knows a practical truth, that it is not too difficult to be living outside in the wild, when it is warm and the countryside of rocky lands beckon.

The nomad sets off again, taking a one way flight to Crete, satisfied that the income he has created will see him through. It will not be enough to house him permanently, but coupled with the skills that he has developed to live in the wild and the money that he has saved from the building work, he aims to be gone for the entire winter. It is December and yet again, the wanderer leaves a dreary winter England and heads for the sun and so it is that Adam is on the road again, jotting in his journal.

He who wanders free may not know the comfort of regularity, but he is sure to meet incredible times ahead, times that will enrich his life forever. When we cease to be undaunted by the torrents of judgement and jealousy that rain upon us when the glow of an imminent journey flows from our being – we step into an entirely different world, an altered state of travel consciousness. A place from which we emanate untold potential joy.

Continued in 'The Adventures of a Greenman - Part 11: Raw Travel Crete'

*

<u>On Wwoofing</u>

Wwoofing brings me closer to all that sustains me. It reminds me how to be close to the elements, work with them and submerge myself in all that is natural. My mind quietens, my skin can breathe – as I move away from the clogged denseness of towns and cities. Back into the natural rhythms of life – in time with the seasons. I am reminded of my basic human needs. I refresh myself by living them – mindfully, as consciously as I can. I surround myself with the simplicity of a facadeless nature. Animals, plants, trees and pure air. I feel the type of satisfaction that can only be felt after a days toil with the soil.

A. Greenman   Summer 2009

Go well on your wwoofing journey.
May your pack be light, and your sleep be tight.

Other titles by A.Greenman:

How to Take a Gap Life - *Sometime a Gap Year is Just Not Enough!*
The Wisdom of Travel - *Words to Inspire*
A. Greenmans Short Stories

The Adventures of a Greenman Series:

Part 1 - *The First 30 Years of a Greenman*
Part 2 - Raw Travel Brazil
Part 3 - Raw Travel France
Part 4 - Raw Travel Spain
Part 5 - Raw Travel India
Part 6 - Raw Travel New Zealand
Part 7 - Raw Travel Wales
Part 8 - Raw Travel Eastern Europe
Part 9 - Raw Travel Hungary
Part 10- Raw Travel England
Part 11 -Raw Travel Crete
Part 12 -Raw Travel Italy
Part 13 -Raw Travel Europe
Part 14 -Around the World in a Decade
Part 15 -I Travel Light:
The Man Who Walked Out of the World.

A.Greenman's Short Stories

Can Blind Girls Travel?
Kate wants to see the world, only she cannot, for she is blind, and seeks to find a way how.

The Nature of London
Omar tires of his daily commute and longs for peace and quiet and to spend more time in his Victorian walled garden.

The Spell of the Word Brand
Brand words come into our lives as softly as the morning dawn, but do we know what power they have on our lives?

...**Bonus Story** *Walking in the Clouds*
A true story based on A.Greenman's own experience of an expedition to Snowdonia National Park, Wales.

How to Take a Gap Life:

Sometimes a Gap Year is Just Not Enough!

Based on a true story. Joseph took some time out and embarked on a Gap Year. 10 years later he found that he had actually taken a Gap Life and was ideally qualified to apply for a job he has seen advertised in a newspaper, for a 'Gap Year' columnist. A humourous and inspiring account and in an in depth CV. Based on A.Greenman's own nomadic wanderings through England, Scotland, Wales, France, Spain, Portugal, Italy, Greece, Hungary, Latvia, Ukraine, Brazil, India and New Zealand, since 1999. Packed with real life adventures, portraying a *Master Gapster's* decade in a fast and addictive tale, led by the character Joseph. He volunteers on organic farms, at orphanages, and works as a carpenter, writer, healer and teacher.

The Adventures of a Greenman:
Raw Travel Series

True Stories

This short story series explores the wanderings of A. Greenman in 15 books, covering England, Wales, France, Spain, Hungary, Italy, Greece, Eastern Europe, Brazil, New Zealand, India, the thirty years prior to his journeys and more...

Part 2

I Travel Light
The Man Who Walked Out of the World
A True Story

A book collating all of A. Greenman's adventures since birth, and even a little before! All in one epic edition. Formed of *Parts 1 to 14* of the *Raw Travel Series*.
Part 15 of The *Adventures of a Greenman* is:

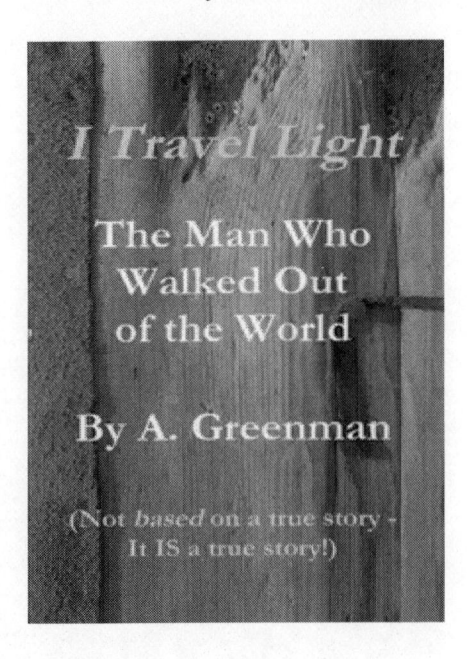

I Travel Light

The Man Who Walked Out of the World

By A. Greenman

(Not *based* on a true story - It IS a true story!)

Praise for the novel

(From someone who really knows what is good for you!)
"When I had something important to do, I found myself reading this book instead - it is BRILLIANT!"

Dr. Longmore, author of the bestselling
'Oxford Handbook of Clinical Medicine'

Preview

I Travel Light

The Man Who Walked out of the World

Prologue

At parties, Adam was often asked 'What do you do?' and he answered 'I travel'. It was a truthful answer because for ten years, that is indeed what he did, travelled. Between 2000-10, the young Englishman wandered the earth, on wild and adventurous travels. Yet how does one become a perpetual wanderer, a professional traveller? In order to answer that question, let us go back to the beginning, the very beginning. It was a time when Adam was choosing where to be born, a place where he may lead a nomad's life. There was only one stipulation, laid down by the great oneness, he must be born to a 'westernised' family, the implication of which was that it would be far harder to lead a nomadic lifestyle, harder that is, in a modern world - as opposed to remote parts of Africa, Asia or even Europe. Nevertheless, this was his lot, this was the way and the path Adam must walk. Take my hand and rest a while and I will take you on a journey, into the world of a wanderer, but please hold on tightly, for we shall barely stop for breath.

* * *

The new spirit looked out over the universe, through the timelessness of his soul, the blue planet caught his eye again. Soon he would have to decide where he would be born, even now he was tempted to choose one of the other planets, perhaps one of the deep green ones again, far away in another space. He would not be human there or know the perils of having a body, yet there were different lessons for the spirit to learn. However, souls were queuing up to go to earth at this time, the beginning of the 1970s. It was a tough course to follow, but the rewards were high. He would have to choose within the next few thoughts, a gap was appearing for him. The fresh spirit had been waiting for the right parents, in order to play out the lessons he must learn on earth.

Gap, he thought.

He was thinking about the recent human habit of taking time out, a gap year and knew that in order to fulfil all his tasks on earth this time, he would have to engineer some sort of extension to the concept of gap years. Sometimes a gap year was just not enough, he would need many, a gap life perhaps. Consecutive gap years, in order to walk the difficult path he had chosen.

He watched the parents which would have this baby, it would be a second boy to follow the one who had already been born. He looked at the setting, a farm in the south of England. They were young strong parents, high spirited and adventurous. The free spirit emptied his mind, cleared any trace of intentions from his thoughts and glowed like a small star, like the millions of others waiting to be born, unseen by most humans. He could go anywhere that his imagination allowed him to - where the thought goes, the energy flows, but it was not a time to travel the many other parts of the Universe that he so enjoyed. His mind must be empty now, just a spark in the ether. He had two thoughts left to use.

Firstly. Shall I go to the green planet or the blue planet?

Secondly. This could be my last incarnation. Earth will not be a smooth path for me, yet it is where I will truly know the nature of the 'free spirit' and complete my lessons.

He was pulled to the blue and must trust the attraction that he felt. There was a moment's hesitation as his spirit prepared for the change, once he had made the move, the young human would not consciously remember anything about the time before he was born for another decade. He would feel a missing, yes, a longing feeling to be back somewhere, but the young one would not know what or where that place was.

It must be earth, I must go there, he decided.

*

Available via:
www.greenmansbooks.com

...and more information on A. Greenman; photos, paperbacks, e-books, book launches, newsletters, public events and signings.

*

Thank you for reading!

About the Author

A. Greenman was born on a farm in England in 1971. Between 2000-10, he wwoofed in France, Spain, Portugal, Hungary, England, Wales and Scotland. He has also participated in voluntary work in Latvia, Brazil, India, and travelled through many other countries. He now lives in Sussex, England where he continues to write both fiction and non-fiction books, in between collecting back packs, future dreams and planning other land based adventures.

Why not write about your wwoofing adventures and forward them to A. Greenman for consideration, to include in future editions of this book? email: a@greenmansbooks.com

Printed in Great Britain
by Amazon.co.uk, Ltd.,
Marston Gate.